QUANTUM MECHANICS THROUGH NOBEL

Revealing Quantum History
from the Nobel Archives

QUANTUM MECHANICS THROUGH NOBEL

Revealing Quantum History from the Nobel Archives

Mats Larsson

Stockholm University, Sweden

World Scientific

NEW JERSEY • LONDON • SINGAPORE • BEIJING • SHANGHAI • HONG KONG • TAIPEI • CHENNAI • TOKYO

Published by

World Scientific Publishing Co. Pte. Ltd.
5 Toh Tuck Link, Singapore 596224
USA office: 27 Warren Street, Suite 401-402, Hackensack, NJ 07601
UK office: 57 Shelton Street, Covent Garden, London WC2H 9HE

British Library Cataloguing-in-Publication Data
A catalogue record for this book is available from the British Library.

QUANTUM MECHANICS THROUGH NOBEL
Revealing Quantum History from the Nobel Archives

ISBN 978-981-98-2898-2 (hardcover)
ISBN 978-981-98-2899-9 (ebook for institutions)
ISBN 978-981-98-2900-2 (ebook for individuals)

For any available supplementary material, please visit
https://www.worldscientific.com/worldscibooks/10.1142/14763#t=suppl

Desk Editor: Rhaimie Wahap

DEDICATION

I dedicate this book to Lars Brink, who was my mentor when I first became a member of the Nobel Committee in 2013, and to my wife Viktoria, who has been my strong supporter during my work with this book.

Photo from the Nobel archive at the Center for History of Science at the Royal Swedish Academy of Sciences. Mats Larsson (left) and Karl Grandin, director of the center. (Courtesy Anne Miche de Malleray).

FOREWORD

In 1964, Richard Feynman (Nobel Laureate 1965) said "I think I can safely say that nobody understands quantum mechanics."[a] Taken literally and in isolation, the meaning is clear — there is no human understanding of quantum mechanics. However, what he meant is that there is no analogous everyday experience that we can draw upon to construct intuitive models. Indeed, the mathematical framework was (and is) so well developed that experimentally tested quantitative predictions abound and form the basis of modern technology, now viewed as banal by the public that uses it.

Despite the stunning conceptual and practical advances, it is an old cliché to say at the beginning of a course that undergraduates "know" they understand quantum mechanics but cannot calculate very well, graduate students question their understanding but are becoming proficient calculators, and professors *know they don't understand* quantum mechanics but are expert calculators. Despite this sardonic perspective, which has many levels of interpretation, it encapsulates the reality that the development of the axiomatic framework that is the stuff of textbooks involved continual practical and interpretive groping in the dark with fits and starts of light. Of course this can be said about any field of basic research, but it is particularly relevant to quantum mechanics; it lurched forward — at once

[a] Feynman, Richard (1965). *The Character of Physical Law*, Modern Library. ISBN 978-0-679-60127-2, which is based on his 1964 Messenger Lectures at Cornell. The quote comes from Lecture 6, Probability and uncertainty — the quantum mechanical view of nature.

dragging along and discarding classical concepts — as experiments and new intuition developed.

The punctuated evolution of quantum mechanics is mirrored in both the process and the people associated with the Nobel Prizes in the field, beginning with the complexities of the 1918 Prize awarded to Max Planck, whose name — through Planck's constant — is inextricably linked to the field, and ending with the 1963 Prize awarded to Eugene Wigner[b]. These pioneers in the development of quantum physics bookend the treatment herein.

A long-standing member of the Nobel Committee for Physics, Mats Larsson is amongst a small number of people with this background who has delved deeply into the archive maintained at the Royal Swedish Academy of Sciences. His efforts will reward the reader on many fronts.

Firstly, despite the amount of public information available, the Nobel Prizes are still cloaked in mystery, largely because, in the main, the archives are in Swedish and thus historians of science have both language and conceptual barriers to overcome, neither of which are impediments to the author. More importantly, however, is the fact that archive documents remain closed for fifty years or until a laureate has passed away. Therefore, since the deliberations of contemporary prizes are secret, each year a prize may appear to have come "out of nowhere" to many observers. Importantly, in the first chapter the mechanics of the process and its relation to the rules of the Nobel Foundation, are laid bare. In this regard, in the remainder of the book, the drama of deliberations, personalities and the then contemporary cutting-edge understanding, or lack thereof, are properly framed through the lens of the regulations that must be followed.

Secondly, the fact that the process involves the entire community, which has obviously grown in the last 125 years, is not commonly appreciated. While the committee receives many hundreds of nominations annually, they are organized and assessed within the framework of a continuous stream of reports solicited from international experts across the subdisciplines. The book demonstrates that one of the most prestigious prizes in the world is the consequence of a collective and collaborative phenomenon.

[b]The reader will learn that the strictures of the prize nearly prevented this chapter from appearing.

Thirdly, the confluence of prospective and retrospective scientific developments and debates involved in the prize provides the student of physics an essential perspective that what we read in our textbooks is the result of a tortuous, hard-won process and interplay between theory and experiment. This reflects the fact that the prizes described provide the skeleton of quantum mechanics constructed in the nonlinear thought processes of humans.

Finally, having worked closely with the author on the committee, I have intimate experience with his serious approach and understanding of how the nuances of history, priority and science writhe through the process. This book is clearly born of a unique passion, emerging from years of grappling with the pantheon of issues shaping the Nobel Prize, from physics to historical precedence and everything between. The reader will have the pleasure of sharing that passion through this compelling journey of the development of quantum mechanics.

John Wettlaufer
Yale University
Nordic Institute for Theoretical Physics

PREFACE

On Thursday, November 9, 1933, the Royal Swedish Academy of Sciences decided to award the 1932 Nobel Prize in Physics to Werner Heisenberg and to divide the 1933 Nobel Prize between Erwin Schrödinger and Paul A.M. Dirac. No physics prize had been awarded since 1930, when Chandrasekhara Venkata Raman received the prize for the discovery of the effect that now bears his name. This was the longest gap in the peacetime history of the Nobel Prize in Physics.

I became aware of this curious fact when my colleague Lars Brink at Chalmers University of Technology — who sadly passed away in 2022 — invited me to Nanyang Technological University (NTU) in Singapore to give an after-dinner speech about the 1932/33 Nobel Prizes, in connection with the conference *90 Years of Quantum Mechanics*, which Lars organized together with colleagues at NTU in early 2017.

At that time, I was entering my third year as a member of the Nobel Committee for Physics, and Lars knew both my deep interest in the history of physics and my familiarity with the Nobel archives. The 1933 awards concerned three of the greatest physicists of the 20th century, and the period between 1925 and 1927 was truly remarkable. It therefore seemed a straightforward task to prepare an engaging after-dinner talk for the conference in Singapore.

After my after-dinner speech in 2017, I gradually expanded the talk into a regular seminar and eventually into a full colloquium. In 2022, I was co-organizer of the first Nobel Symposium ever held outside Scandinavia, at the Stellenbosch Institute for Advanced Study in South Africa. When a last-minute cancellation occurred, I stepped in to give a presentation on the 1932/33 Nobel Prizes in Physics.

The talk had been prepared for a physics audience, but it was scheduled on the morning of the second day — the official opening of the Symposium — when many guests in attendance were not physicists. To my surprise, I received numerous positive comments from non-physicists, which was, of course, very encouraging. If the topic could engage such a diverse audience, perhaps a book on the subject might also reach beyond the physics community.

Needless to say, this is not a textbook on quantum mechanics. My aim is not to teach the theory, but to trace the history of how it was recognized and rewarded. Whenever a brief reference to the underlying physics may help the reader follow the story, I will point to standard textbooks for further detail.

Much of the material presented in this book is drawn from the Nobel archive, which provides access to original nominations, committee reports and correspondence. These documents, preserved in the archives of the Royal Swedish Academy of Sciences, offer a unique perspective on the deliberations behind each award. Whenever possible, I have complemented these primary sources with contemporary scientific papers, Nobel lectures, and relevant biographical and historical studies. References are provided in a consistent format at the end of the book, allowing the interested reader to trace the material in greater detail.

Mats Larsson

CONTENTS

Chapter 1

HOW A NOBEL PRIZE IN PHYSICS IS AWARDED

1.1 Alfred Nobel's Will and the Origin of the Prizes

The central document is the will that Alfred Nobel wrote on November 27, 1895, at the Swedish-Norwegian Club in Paris. It was the third and final version of his will, and it was handwritten in Swedish. In the English translation (see https://www.nobelprize.org/alfred-nobel/alfred-nobels-will/), the will starts as follows: "All of my remaining realisible assets are to be disbursed as follows: the capital, converted to safe securities by my executors, is to constitute a fund, the interest on which is to be distributed annually as prizes to those who, during the preceding year, have conferred the greatest benefit to humankind." This is the umbrella sentence applying to all the five prizes.

It then continues: "The interest is to be divided into five equal parts and distributed as follows: one part to the person who made the most important discovery or invention in the field of physics; one part to the person who made the most important chemical discovery or improvement; one part to the person who made the most important discovery within the domain of physiology or medicine; one part to the person who, in the field of literature, produced the most outstanding work in an idealistic direction; and one part to the person who has done the most or best to advance fellowship among nations, the abolition or reduction of standing armies, and the establishment and promotion of peace congresses."

The phrase concerning physics — "who has made the most important discovery or invention" — has occupied every Nobel Committee for Physics since the beginning.

Alfred Nobel also pointed to the institutions that should award the prizes — in the case of physics and chemistry, the Royal Swedish Academy of Sciences in Stockholm.

What is striking about the will is that no fund existed when Nobel wrote his will; in other words, no beneficiary existed. The dramatic events that followed after Alfred Nobel passed away on December 10, 1896, are described in detail in Ingrid Carlberg's biography (2019, 2023). The story reads like a detective novel, with Ragnar Sohlman, Alfred Nobel's collaborator, as executor of the will. By the year 1900, the Nobel Foundation, a private institution, had been formed, and the first Nobel Prizes were awarded in 1901. The main mission of the Nobel Foundation is to manage Alfred Nobel's fortune. The prize-awarding institutions are guaranteed independence in their work of selecting recipients.

1.2 Statutes of the Nobel Foundation

The Nobel Foundation also had the important task of deriving more detailed explanations and instructions for the prize-awarding institutions than those available in the will. Statutes of the Nobel Foundation were approved by the Crown on June 29, 1900. The statutes were amended for the Royal Swedish Academy of Sciences on June 29, 1973, with approval from the Government of Sweden. The statutes for physics and chemistry are the same, with the only difference attributable to Alfred Nobel himself. For both physics and chemistry, a person can be awarded a prize for discoveries, but for physics, inventions can also be awarded, whereas for chemistry one can instead be awarded for improvements — a small but important difference. In the rest of this book, the term "prize" refers to the Nobel Prize for Physics. If another prize is mentioned, it will be clearly stated which of the other science prizes is concerned.

According to §4, the following holds: "A prize may be equally divided between two works, each of which is considered to merit a prize. If a work that is being rewarded has been produced by two or three persons, the prize shall be awarded to them jointly. In no case may a prize amount be divided between more than three persons."

This paragraph does not require that a divided prize have any connection between the two parts, only that each part merit a prize. Sometimes, there has certainly been a strong connection, such as in the 1903 prize, with one half going to Henri Becquerel and the other half jointly to Marie and Pierre Curie. Becquerel was awarded the prize for his discovery of spontaneous radioactivity, and he is explicitly named in the citation for the prize to the Curies. The 1903 prize was an unusually complicated affair, but it would lead too far astray to discuss it further; the reader is referred to Grandin (2020) for an authoritative and detailed account.

As we will see when we discuss the 1954 prize, divided between Max Born and Walter Boethe, the connection was very weak.

Thus, the maximum number of persons who can receive a given year's prize is three, and the maximum number of citations is two.

§5 reads: "A work may not be awarded a prize, unless it by experience or expert scrutiny has been found to be of such outstanding importance as is manifestly intended by the will. If none of the works under consideration is found to be of the importance indicated in the first paragraph, the prize money shall be reserved until the following year. If, even then, the prize cannot be awarded, the amount shall be added to the Foundation's restricted funds."

As we shall see later, this paragraph has been used several times during peacetime (the war times 1914–1918 and 1939–1945 will not be discussed), in particular in connection with the prizes in 1932/33.

The phrase in the will, "during the preceding year," was not possible to uphold when the Nobel Foundation formulated the statutes. Instead, §2 states, "The provision in the will that the annual award of prizes shall be intended for works 'during the preceding year' should be understood in the sense that the awards shall be made for the most recent achievements in the fields of culture referred to in the will and for older works only if their significance has not become apparent until recently."

Closely related to §2 is §3, which defines the concept of "work": "To be eligible to be considered for a prize, a written work shall have been issued in print or have been published in another form, to be decided on its own behalf by each prize-awarding body."

It is obvious that §2 gives a lot of leverage to the Nobel Committee, and taking a look at the history of the prize, it is clear that the committee has used this leverage. There are also cases when it has been possible to meet Nobel's original criterion. A recent example is the 2017 Nobel Prize awarded to Rainer Weiss, Barry Barish and Kip Thorne "for decisive contributions to the LIGO detector and the observation of gravitational waves." The first observation was published in *Physical Review Letters* on February 11, 2016, and the key persons were awarded the prize in 2017. There were even speculations that the award could come in 2016, but this was based on a misunderstanding of the statutes. Even though the gravitational wave signal was observed on September 14, 2015, there has to be a written work. The formulation of the phrase "the preceding year" makes it impossible to award a discovery the same year it was published.

§4 stipulates that "Work produced by a person since deceased shall not be considered for an award. If, however, a prizewinner dies before he has received the prize, then the prize may be presented."

When one goes through the Nobel archive, the 1910 prize may appear puzzling with respect to §4. The Swedish physicist Knut Ångström was nominated by Wilhelm Röntgen in 1910 for his work on the radiation of heat from the sun. Knut Ångström was a member of the Nobel Committee from 1900 until 1909, and the nomination by Röntgen was the first and only one he received. Gustaf Granqvist, who had joined the committee in 1904, wrote a special report about Ångström's work in which he concluded that the work was of Nobel caliber. However, Knut Ångström died on March 4, 1910, so one can wonder why the Nobel Committee continued to consider Ångström even after he had passed away. In fact, in the committee's report to the Royal Swedish Academy of Sciences, dated September 26, 1910, Knut Ångström was still the first name. The reason for this is the following: the original formulation of §4 from the year 1900 reads, "The work of any person since deceased cannot be submitted for award; should, however, the death of the individual in question have occurred subsequent to a recommendation having been made in due course for his work to receive a prize, such prize may be awarded." The nomination by Wilhelm Röntgen was received prior to Knut Ångström's demise, so the Nobel Committee felt that it was according to the statutes to award him. However, the committee was not quite confident that the Academy *in pleno* would follow their suggestion and put forward a second name, Johannes Diderik van der Waals. The Academy decided to award van der Waals, but we do not know the rationale for this. Discussions or voting figures are according to the statutes not recorded at the final plenary of the

Academy when awarding the prize; only the decision is preserved in the minutes.

The year 1910 is noteworthy because of the campaign orchestrated by the Swedish mathematician Gösta Mittag-Leffler for an award to the French mathematician Henri Poincaré. The campaign resulted in 34 nominations for Poincaré, of which 15 came from France, with the nominators distributed across Japan, Russia, several European countries, and the US. Curiously, Mittag-Leffler himself did not nominate Poincaré that particular year, although he always had, in his capacity as a member of the Royal Swedish Academy of Sciences, the right to nominate. The 34 nominations stood as a record for a long time and were almost successful. Whereas the Nobel Committee was unanimous in putting forward Ångström as the prime candidate, the committee was split concerning van der Waals; two of the five committee members favored Poincaré.

In 1974, a provision to §10 in the statutes was added, stating that "[a] prize-awarding body may, after due consideration in each individual case, permit access to material which formed the basis for the evaluation and decision concerning a prize, for purposes of research in intellectual history. Such permission may not, however, be granted until at least 50 years have elapsed after the date on which the decision in question was made."

Obviously, without this provision, this book could not have been written.

In August 1981, a Nobel Symposium on the theme "Science, Technology and Society in the Time of Alfred Nobel" was organized in Karlskoga, Sweden, where Alfred Nobel spent the last three years of his life. The proceedings were published with the same title by Pergamon Press, Oxford, in 1982. At the symposium, some individuals who had been granted permission to use archival materials for research presented papers on Nobel Prizes between 1901 and 1930. This symposium is, to this day, the only one dedicated to Nobel Prizes. One can note that most of the material presented in this book was not open for research in 1981.

1.3 The Procedure for Awarding a Nobel Prize

The procedure for awarding a Nobel Prize has been almost unchanged since the beginning. It rests on three legs. The first leg is the nomination of candidates by scientists with the necessary competence. A few categories have permanent nomination rights, such as members of the Royal

Swedish Academy of Sciences (including foreign members), members of the Nobel Committee, scientists who have been awarded the prize, and professors and assistant professors of physics in Sweden, Denmark, Norway, Finland, and Iceland. Then, there are holders of corresponding chairs from at least six universities selected by the Academy with a view to ensuring the appropriate distribution of the commission over the different countries and their seats of learning. Finally, there are other scientists whom the Academy may see fit to invite proposals.

The permanent nomination rights are more than a mere technicality. Marie Curie shared one half of the 1903 Nobel Prize in Physics with her husband, Pierre, yet she had not been formally nominated during 1903. This might appear to contradict the statutes, but the explanation is as follows: Marie and Pierre Curie were nominated by Charles Bouchard in 1902; however, his letter arrived on February 4 — after the official deadline of January 31. Bouchard was a foreign member of the Royal Swedish Academy of Sciences and therefore held permanent nomination rights. According to the statutes, his late letter was counted as an early nomination for 1903 (Grandin, 2020).

In the first year, 1901, there were 30 nominations in physics, of which 16 were for Wilhelm Röntgen. These days, there are about 250–350 names that are nominated during a specific year, according to nobelprize.org.

The second leg is the special reports submitted to the Nobel Committee. For many years, these special reports were written by committee members, but from 1964 onward, special reports by non-committee scientists were commissioned by the committee. Most likely, some readers of this book have sent special reports to the Nobel Committee. They are, according to §10, kept secret for 50 years.

The third leg is a report written by the Nobel Committee and submitted to the Royal Swedish Academy of Sciences. The proposal from the Nobel Committee is discussed in the Academy's physics section (Class III, the Physics Class), which will vote on whether to support the committee's proposal. During the early years, the physics class was not much larger than the committee, whereas these days the class is much larger than the committee. Often, several previous members of the committee participate in the meeting. The reason for the change in size is that new members can join the Physics Class when a member turns 65, whereas earlier, a new member could only join after somebody died.

Finally, on the day of the decision, the Royal Swedish Academy of Sciences meets in a plenary session open to any member elect. No discussions or votes are recorded in the minutes — only the decision. The proposal from the Nobel Committee is presented, which can be a concrete award to one, two or three persons, but it can also be a proposal to postpone the prize to the next year. After the committee has presented its proposal, the opinion of the Physics Class is presented. The floor is then open for any member of the Academy to propose another award, including postponing. However, these discussions cannot be extracted from the minutes. On several occasions, these plenary discussions have played a decisive role in shaping the final outcome, sometimes even overturning the recommendation of the Nobel Committee and the support of the Physics Class.

Fig. 1.1 A drawing of Wilhelm Röntgen by an unknown artist published in the national Swedish newspaper *Aftonbladet* on December 9, 1901. In the early years, December 10 was not only the day for the prize ceremony but also the day when the laureates were made public. The newspapers had no problem figuring out who the winners would be, and in 1901 *Aftonbladet* were 100% correct concerning the four prizes to be awarded in Stockholm (physics, chemistry, medicine or physiology, and literature). In 1910, the decision date was separated from the award ceremony day of December 10, and this is still the procedure. Used with permission.

Once the decision has been taken, it is final and cannot be appealed. The names of the Nobel Laureates are then announced. The Nobel Prize award ceremony takes place in Stockholm on December 10, where the Nobel Prize laureates receive their Nobel Prize medal, diploma and a document confirming the prize amount. They are also requested to give a public lecture about the work for which they have been awarded. These days, this lecture is given on December 8, but for the prizes discussed in this book, the Nobel lectures were given a day or two after the award ceremony.

Chapter 2

THE EARLY QUANTUM MECHANICS NOBEL PRIZES, 1918, 1921 AND 1922

2.1 The Nobel Prize in 1918 to Max Planck

Carlberg (2019, 2023) describes how some people in Sweden took the formulation in Nobel's will — "during the preceding year" — too literally and did not realize that in the Statutes of the Nobel Foundation, the wording is less strict. Thus, they believed that Max Planck would be the natural choice because of his radiation law, formulated in 1900. However, Planck was not nominated in 1901, and Wilhelm Röntgen received the vast majority of nominations and became the first laureate in physics.

Bengt Nagel (1982), secretary of the Nobel Committee from 1974 to 1988, provided a detailed account of how Max Planck was finally awarded the 1918 Nobel Prize in 1919. This will not be repeated here; instead, we will focus on the most dramatic events during the almost twenty years it took before Planck received the award — the events of 1908.

Before returning to the archives, it is worth taking a look at the situation regarding radiation laws in the second half of the 19th century. In 1860, Gustav Kirchoff empirically established that a black-body spectrum should look like the curves in Figure 2.1. The challenge for physicists was now to find a function that would correctly predict energy density at all wavelengths.

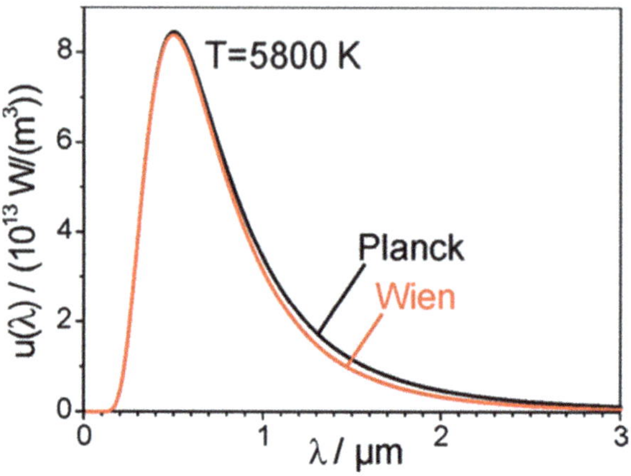

Fig. 2.1 The energy density emitted by a black body as a function of wavelength. The red curve is the function developed by Wilhelm Wien, and the black one is the Planck curve. Wien correctly predicted how the peak of the curve would change with temperature (Wien's displacement law). Wien was awarded the 1911 Nobel Prize in Physics for his discoveries regarding the laws governing the radiation of heat.

In 1879 and 1884, Josef Stefan and Ludwig Boltzmann noted that the area under the curve was proportional to the fourth power of temperature (T^4), which became known as the Stefan–Boltzmann law. In 1893, Wilhelm Wien used thermodynamics to derive that the wavelength at which the curve peaked was proportional to the inverse of the temperature (Wien's displacement law). Three years later, he used the statistical mechanics developed by Boltzmann to derive the first function that provided a good fit to the experimental data.

Experimental physicists played a very important role in the development of black-body radiation physics, so it is appropriate to mention them here: Otto Lummer, Ernst Pringsheim, Heinrich Rubens, and Ferdinand Kurlbaum. Lummer was nominated in 1910 and 1911, Rubens in 1907, and twice in 1909, whereas Pringsheim and Kurlbaum were never nominated. It was their careful measurements of the black-body radiation spectrum that made it apparent that Wien's radiation law did not agree with the experimental results at long wavelengths.

In October 1900, Planck used thermodynamics to derive a new radiation law, which agreed very well with the experimental results. But what was the physical meaning? Planck was sceptical of Bolzmann's statistical mechanics, but having exhausted the theories of electromagnetism and thermodynamics, he finally gave up and resorted to Boltzmann. Planck found that he could derive his radiation law by assuming that the oscillators that gave rise to black-body radiation oscillated with discrete energies equal to $h\upsilon$, where υ is the frequency of the radiation and h became known as Planck's constant. Since 2019, this constant has had a fixed value of $6.62607015 \times 10^{-34}$ J·Hz^{-1}. For more details about Planck's introduction of the quantum hypothesis, which is beyond the scope of this book, the reader is referred to Nauenberg (2016).

Planck was first nominated in 1907, and the following year, in 1908, he was — without his knowledge, of course — involved in one of the most dramatic events in the history of the physics prize (with a cut-off set at the year 1973 and a few cases earlier where the laureate was still alive and the archive must remain closed). Planck received a single nomination in 1908 submitted by the Swedish mathematician Ivar Fredholm (1866–1927), who proposed that the prize be divided between Planck and Wien. They competed with Gabriel Lippmann, whom the Nobel Committee had already declared worthy of a prize for his method of reproducing colors photographically. The Committee prepared separate reports, one of which suggested Lippmann. It is clear from the minutes of a committee meeting on September 18, 1908, that the members were reluctant about the Planck alternative, with all members except Arrhenius putting their names in the minutes. The Committee solicited a special report from Committee member Svante Arrhenius, who concluded that Planck should receive an undivided prize. This is also what the Nobel Committee, after the greatest hesitation, put forward in their report of September 24, 1908, to the Academy. At the Nobel discussion in the Physics Class, in which nine members participated (five from the Nobel Committee and four others), only one was against.

We do not know any details of what happened at the meeting in the Academy in November 1908, except for the decision to award the prize to Lippmann. In his article, Nagel (1982) used a letter from Mittag-Leffler to P. Painlevé and a letter from Fredholm to Mittag-Leffler to conclude that the two mathematicians could accept a divided prize for Planck and Wien, but not an undivided prize for Planck. However, one may speculate that with four members of the Nobel Committee reluctant and the hostility between Arrhenius and Mittag-Leffler, the Academy preferred the safer alternative, Lippmann.

Despite nine nominations for Planck in 1909, the Committee was still not prepared to recommend him, suffering the scars of the 1908 battle.

In 1911 the Committee put forward Wien for an undivided prize for his radiation displacement law, which has a solid theoretical foundation. It is also easier to measure the peak of a radiation curve than to measure its shape. In a presentation speech on December 10, 1911, the Swedish National Librarian Erik Dahlgren mentioned every important person in radiation physics, and in the written account, the names are given in capital letters in Les Prix Nobel 1911 (1912).

It would take until 1919 for Planck to receive his Nobel Prize, reserved from 1918. It was a gradual process for the Nobel Committee to finally conclude that the time was ripe for awarding Planck. Arrhenius wrote a special report in 1919 in which he stated, "All these circumstances make it evident that the quantum theory of Planck is the leading theory in the modern development of physics; although it has not yet obtained a final, satisfactory formulation, quantum theory plays a role making it fully comparable even with the foremost theories so far invented by the pioneers of this science" (Nagel 1982).

The Committee noted that no other scientist had been nominated by so many different and such competent nominators. In its report to the Academy, the Committee wrote, "It has become increasingly probable that the Planck constant, the discovery of which summarizes the theory, corresponds to some reality in the constitution of matter, and there is no longer any doubt but that this constant gives the measure of an atomistic property, albeit of yet unknown nature" (Nagel 1982).

The Committee's suggestion for the 1918 prize met no objections in the Physics Class or in the Academy (Nagel 1982).

The nominations of Planck in 1919 are noteworthy because this was the very first time Albert Einstein and Max Born made nominations. In the English translation of Abraham Pais[a] (Pais 1982), Einstein writes, "Because of this work [two articles by Planck in *Drudes Annalen* from 1901], the author has not only given a first exact determination of the absolute size of atoms but especially [he has] also laid the foundation for

---

[a]Pais writes on p. 513 in his book that Einstein's nomination is "undated but no doubt written in the fall of 1918." It is correct that the nomination letter is undated, and it reached the Nobel Committee in Stockholm on October 18, 1918. What is peculiar is that Einstein did not wait for the Academy's decision, which came in November 1918 (postponing the physics prize to 1919). As we will see, this also happened with Einstein's later nominations.

the quantum theory, the fertility of which for all of physics has become manifest in recent years."

When the Academy announced the Nobel Prizes in Physics and Chemistry in November 1919, three German scientists were awarded: Max Planck received the 1918 prize in physics, Johannes Stark the 1919 prize in physics, and Fritz Haber the 1918 prize in chemistry. No prize in chemistry was awarded in 1919.

In the presentation speech on June 1, 1920 (Les Prix Nobel 1919–1920), given by the president of the Academy, Åke Ekstrand, both the law of the photoelectric effect as formulated by Einstein in 1905 and Niels Bohr's work on atomic structure in 1913 were mentioned as strong support for Planck's theory of energy quanta. As we will see in the next chapter, it would take only a few years before Einstein and Bohr, on the very same day, were awarded the 1921 and 1922 Nobel Prizes in Physics.

The peculiar date requires some explanation. The idea was that instead of holding the prize ceremony in the dark and cold of December, a June celebration would be far more inviting. The First World War had prevented the Nobel Prize ceremonies from being held for more than six years. Heike Kamerlingh Onnes was the last physicist to come to Stockholm, in 1913, before the war made it impossible for laureates to travel.

As a result, many Nobel lectures were given in the days following June 1: Max Planck on June 2, Johannes Stark on June 3, Max von Laue (1914 Prize) on June 3, and Charles Barkla (1917 Prize) also on June 3. Lawrence Bragg, who shared the 1915 Nobel Prize with his father, William Bragg, did not deliver his lecture until September 6, 1922. His father never gave a Nobel lecture.

So why, then, was the June 1 ceremony a one-time event, never to be repeated? Crown Princess Margareta had died on May 1, 1920, at the age of 38, leaving her husband (the future King Gustav VI Adolf) and five children, and the country was in mourning. No members of the Royal Family attended the ceremony — something that had not happened since 1907, when King Oscar II died on December 8, only two days before the scheduled Nobel ceremony.

After 1920, the Nobel Prize ceremonies permanently returned to their traditional date in December, coinciding with the anniversary of Alfred Nobel's death on December 10.

2.2 The Nobel Prizes in 1921 and 1922 for Albert Einstein and Niels Bohr

Already in the presentation speech on June 1, 1920, one gets a hint of how the Nobel Committee would assess contributions to atomic theory: (In my translation from Swedish to English) "Even greater triumphs have been celebrated by Planck's theory in spectral analysis, where Bohr's fundamental work and the complementary works of Sommerfeld, Epstein and others have provided an explanation for the mysterious laws that prevail in this part of science." In fact, Sommerfeld was never close to being awarded a Nobel Prize.

Neither Bohr nor Einstein was close to a Nobel Prize in the way Planck was in 1908, before they actually received the prize. Since Einstein's prize has been discussed in great detail by Grandin (2021) and Bohr's prize has been discussed by Pais (1991), the focus will be on the period 1921–22, when the real action took place.

The person who played a pivotal role in the awards of Einstein and Bohr was Carl Wilhelm Oseen. He earned a PhD in mathematics in 1901 from Lund University and then shifted his interest to mathematical physics, in particular hydrodynamics. He became a full professor in mechanics and mathematical physics at Uppsala University in 1909. He was elected to the Royal Swedish Academy of Sciences and its Mathematics Class in 1921. More importantly, he became an adjoint member of the Nobel Committee for Physics in the fall of 1921. The year after, in 1922, he was transferred to the Physics Class and became a full member of the Nobel Committee, replacing Bernard Hasselberg, who had been a member of the Committee since the beginning in 1900, but passed away in May 1922. More information on Oseen, taken from Karl Grandin's PhD thesis (Grandin 2000), is available in Appendix 1.

Oseen realized that the Nobel Committee was in a deadlock concerning Einstein. Figure 2.2 shows the number of nominations received by Einstein from 1910 to 1922. The number increased significantly in 1921 and 1922, when the results of the total solar eclipse in May 1919 proved Einstein's general theory of relativity correct. However, Committee member Alva Gullstrand, Nobel laureate in medicine or physiology in 1911, was not convinced. In two special reports in 1921 and 1922, he rejected the bending of light and the perihelion motion of

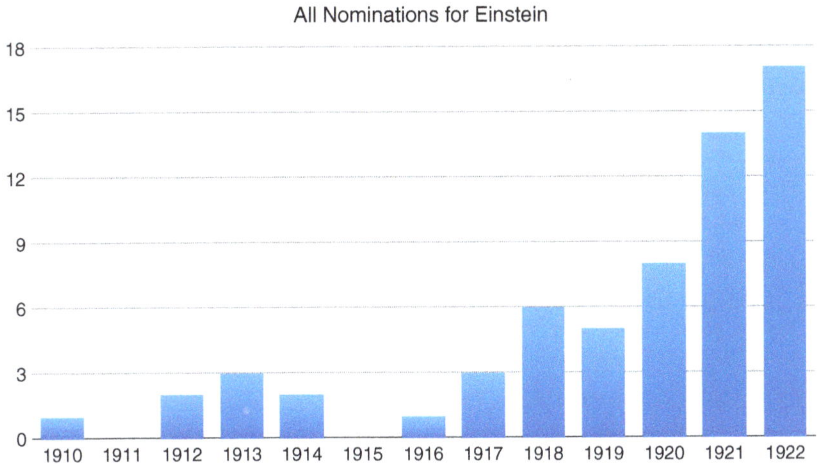

Fig. 2.2 The number of nominations of Albert Einstein 1910 to 1922. Reproduced with permission of Karl Grandin.

Mercury as proofs of the general theory of relativity. At the same time, Einstein received the largest number of nominations, 17, followed by Bohr's 11. Gullstrand's special report convinced Oseen that it would be impossible to award Einstein based on theories of relativity. Instead, Oseen nominated Einstein for his explanation of the photoelectric effect in 1905. In fact, Oseen was the only nominator to nominate Einstein for the photoelectric effect.

The physics prize for 1921 was postponed for one year, which means that the Nobel Committee had two prizes to award in 1922. As it happened, the Chemistry Committee was in the same situation, as will be apparent in a little while.

The Physics Committee commissioned two special reports by Oseen: one for Bohr and one for Einstein. This gave Oseen the opportunity to couple Bohr and Einstein closely, since the photoelectric effect is related to Bohr's atomic theory of 1913. Once the theory of relativity was off the table, Gullstrand and the Nobel Committee accepted Einstein. Oseen's original special report on Einstein in Swedish is given in Appendix 2.

The Physics Class and the Academy *in pleno* had no objections, although the Academy amended the citation proposed by the Nobel Committee.

In Swedish, the citation for Einstein proposed by the Nobel Committee reads, "för hans förtjänster om den teoretiska fysiken, särskildt hans upptäckt av lagen för den fotoelektriska effekten" (old Swedish spelling), which translates to "for his services to Theoretical Physics, and especially for his discovery of the law of the photoelectric effect." The Academy was eager to keep any trace of the theories of relativity out of the motivation and added, "regardless of the value that, after any confirmation, could be attributed to the theories of relativity and gravity, [...] award the 1921 prize [...] to Albert Einstein for his services to Theoretical Physics, and especially his discovery of the law of the photoelectric effect." This text also made it into Einstein's Nobel diploma. This diploma is the only one for which it is stated what the laurate was *not* awarded for.

On December 10, 1922, Arrhenius gave the presentation speeches for both Einstein and Bohr. Einstein was on his way to Japan and was not present, while Bohr was. Arrhenius was careful not to spend time on the theories of relativity. His presentation speech is given at *nobelprize.org* in English translation, but it was given in Swedish, as was the custom until only a few years ago. The presentation speech in Swedish is given in *Les Prix Nobel 1921–1922* (1923). Reading aloud in Swedish the part that deals with the theories of relativity, at the pace I have used when giving presentation speeches in the Stockholm Concert Hall, shows that Arrhenius devoted 35 seconds of his 4-minute speech to relativity. It would take almost 100 years before the general theory of relativity once again was mentioned in a presentation speech — this time in connection with the award to Rainer Weiss, Barry Barrish and Kip Thorne in 2017 "for decisive contributions to the LIGO detector and the observation of gravitational waves." The lion's share of Arrhenius' presentation speech was given to the photoelectric effect, although Einstein's work on Brownian motion was also discussed.

Arrhenius' presentation speech for Bohr was straightforward. He started with Kirchhoff and Bunsen in 1860 and Balmer in 1885, mentioning a number of other spectroscopists of the 19th century. Then he arrived at Bohr via the contributions of Planck and Einstein. Bohr was present at the award ceremony, and we can assume with confidence that he had no problems following the speech in Swedish.

In those days, the compulsory Nobel lecture was given one or two days after the award ceremony of December 10. Bohr gave his lecture in his native Danish language, with the title "Om atomernes bygning" ("The

DAGENS NYHETER.

NOBELPRISET I LITTERATUR ÅT SPANJOR.
EINSTEIN OCH N. BOHR FYSIKPRISTAGARE.
PRISEN I KEMI ÅT TVENNE ENGELSMÄN.

Dramaturgen J. Benavente litteraturpristagare.

Akademins val ägnat att väcka debatt.

Einstein prisbelönad, men ej för relativitetsteorin.

Forskare i atomernas byggnad få övriga tre pris.

Fig. 2.3 The front page of the Swedish newspaper *Dagens Nyheter* (Today's News) on November 10, 1922, the day after the Academy had voted for awarding the prize reserved for 1921 to Einstein and the 1922 prize to Bohr. The text above the photo of Einstein says that Einstein has been awarded the Nobel Prize, but not for the theory of relativity. Thus, the reader is informed by *Dagens Nyheter* about what Einstein was *not* awarded the prize for. Above the photo of Bohr, it is written that three researchers concerned with the building of atoms have been awarded, one of them Bohr and two others, Fredric Soddy, who received the reserved 1921 prize in chemistry, and Francis W. Aston, the 1922 prize in chemistry, for their work on isotopes. Used with permission.

structure of the atom" — English translation of the lecture available at *nobelprize.org*).

Einstein was unable to deliver his Nobel lecture during the award ceremony because of his journey to Japan. Instead, Arrhenius suggested that Einstein come to Gothenburg the following summer to deliver his Nobel lecture. The proposal was not without controversy. Some members of the Academy argued that the lecture should take place in Stockholm and that the chosen date — July 11, 1923 — exceeded the six-month limit stipulated for Nobel lectures. The timing was, however, well chosen. Gothenburg celebrated its 300-year anniversary with a Jubilee Exhibition, and the Swedish king was present for Einstein's lecture, sitting in the middle with Arrhenius on the side. Einstein, for his part, seized the opportunity for a kind of "revenge." He gave his talk in German, which, together with an English translation, is available at *nobelprize.org*. The title in English is "Fundamental ideas and problems of the theory of relativity". This created a third controversy — Einstein did not speak about the topic for which he had been awarded the prize. The photoelectric effect was not mentioned at all. Judging from photographs of the crowded lecture hall, however, it seems clear that the theory of relativity was far more captivating to the audience than the photoelectric effect could ever have been.

Almost 100 years later, in June 2023, my colleagues in Gothenburg organized a public event at the very same park where Einstein gave his lecture with lectures by, among others, Nobel Laureates David Wineland (2012), Serge Haroche (2012), Donna Strickland (2018) and Didier Queloz (2019).

Einstein was never again nominated, and since nobody else could be considered for a Nobel Prize for the theories of relativity, one of the two great physical theories of the 20th century was never awarded. As we have seen and will see later in this book, quantum mechanics has been treated far more generously.

PORTRAIT COLLECTION

In the old physics building of Stockholm University, there was a portrait collection in the lecture hall of Nobel laureates in physics. In 2001, the Physics Department moved to a new building, AlbaNova University Center together with other departments of Stockholm University and the Royal Institute of Technology. The portrait collection was forgotten. In 2017, when I served part time as director of AlbaNova, there was flooding in the basement of the building. When I inspected the flooding, I came across a couple of boxes and "rediscovered" the portrait collection. It was in terrible shape, with most of the frames broken. A staff member restored the collection and put it up on a wall in the AlbaNova building in a discrete place. When my tenure as AlbaNova director ended, it did not take a long time before people insisted that it should be removed; the collections consisted of almost only men. Somebody called it the hall of shame. The collection was removed and put in boxes, this time very gently. I removed ten portraits and had them scanned by a company since I do not have a scanner that could manage portraits.

It was Gösta Ekspong (1922–2017), a member of the Nobel Committee for Physics (1975–1988), who was responsible for the collection. Sadly, he passed away only a few months before I "rediscovered" it. Some of the photos are the same as those found on nobelprize.org, while others differ. Max Planck's portrait is entirely different; Albert Einstein's is similar but not identical. Those of Niels Bohr and Louis de Broglie appear identical, whereas Werner Heisenberg's is completely different and Erwin Schrödinger's slightly so. The portraits of Paul Dirac and Wolfgang Pauli are identical; Max Born's is entirely different, while Eugene Wigner's is identical. Notably, the original citation in Swedish is printed beneath each portrait.

My thanks go to Per Carlson (member of the Nobel Committee 1999–2007), who helped me with the history of the portrait collection.

1921
Einstein, Albert
1879 - 1955
Tyskland
"För hans förtjänster om den teoretiska fysiken, särskilt hans upptäckt av lagen för den fotoelektriska effekten".

1922

1918
Planck, Max Karl Ernst Ludwig
1858 - 1947
Tyskland
"Såsom ett erkännande av den förtjänst han genom upptäckten av elementar-
kvanta inlagt om fysikens utveckling".

Chapter 3

THE NEW QUANTUM MECHANICS AND THE NOBEL PRIZE IN 1929

3.1 The Nobel Prize in 1929 to Louis de Broglie

In 1923, Arthur Compton discovered and theoretically explained what became known as the Compton effect. Classical wave phenomena theories cannot explain the behavior of X-rays scattered by electrons. Instead, Compton used particle-like momentum for light quanta and special relativity to obtain perfect agreement between experiment and theory. If there had been doubts about Einstein's 1905 photon concept, these doubts evaporated after the discovery of the Compton effect.

Louis de Broglie, the main figure of this chapter, received his two first nominations from Orest Khvolson in the USSR in 1925 and 1926. It should be noted that Khvolson was no newcomer to the Nobel Committee; he had already been an active nominator during the period of the Russian Empire, as shown by his nominations of Planck, Poincaré, Einstein and Bohr in 1909, 1910, 1914 and 1917. Thus, he was a physicist whom the Committee had to take seriously. So, what had de Broglie done that motivated these nominations? In his PhD thesis of 1924, "Recherches sur la théorie des quanta" ("Research on the Theory of Quanta"), Louis de Broglie ascribed a wave-like property to a material particle, such as an electron or atom, and derived the famous relation $\lambda = h/p$, where λ is the wavelength of the particle, h is Planck's constant and p is the momentum of the particle (see Griffiths 2005, p. 19). To ascribe a wave-like property to a material particle was an exceptionally bold hypothesis, as there was not a single experimental observation supporting this idea.

Oseen took the 1925 nomination seriously and submitted a special report to the Nobel Committee with a harsh statement (in English translation): "Against the manner in which L. de Broglie has carried out his theory, very serious objections can be raised. However, since the theory has so far had no significance for physics, it seems to me unnecessary to give any account of it."

The discovery of the Compton effect, for which Compton was solely responsible, was an obvious Nobel Prize contender, and the Nobel Committee was ready as early as 1927 for an award. The Committee commissioned a special report by Oseen the same year, not about the Compton effect per se, since its importance and priority were clear, but rather about the theoretical meaning of the Compton effect. This gave Oseen the opportunity to discuss the new atomic theory, although neither Werner Heisenberg nor Erwin Schrödinger yet had been nominated, but there was now a connection between de Broglie and Schrödinger, the latter clearly inspired by de Broglie's wave hypothesis. Oseen now admitted that the second part of his statement in 1925 was no longer valid. He wrote, "The idea that L. de Broglie presented in an often clumsy and dilettantish manner has, through Schrödinger's investigations, acquired fundamental significance for atomic physics."

Yet, Oseen's conclusion about de Broglie, after a quite lengthy discussion, was that "de Broglie's theory can, under these circumstances, hardly be regarded as tenable."

In 1928, Khvolson again nominated de Broglie, but Oseen was unperturbed in his special report and had not changed his mind. Thus, by 1928, the only three nominations de Broglie had obtained came from Khvolson in Leningrad, USSR. Things would take a dramatic turn the following year.

In 1927, Clinton Davisson and Lester Halbert Germer at Bell Telephone Laboratories in New York published a short article in *Nature* about their work on scattering electrons by a single crystal of nickel, followed in December 1927 by a 35-page article in *Physical Review* giving a full description of the experiment (Davisson and Germer, 1927a, b). They observed diffraction patterns that unambiguously revealed the wave nature of electrons.

The paper in *Physical Review* came too late in 1927, one can assume, to be recognized by the community, and it was a long paper full of experimental details. But in 1929 the full impact of the paper was understood. This resulted in 12 nominations for de Broglie, but also nominations for

Davisson (5) and Germer (1). In the special report written for the Nobel Committee by Oseen, 26 pages in length, it was absolutely clear to him that the Davisson and Germer experiment had not only qualitatively but also quantitatively verified de Broglie's formula. Figure 3.1 shows the first page of Oseen's special report, dated March 16, 1929.

A Nobel Prize divided between theory and experiment seemed like a natural solution, and Oseen discussed it. There was, however, a caveat. George Paget Thomson at the University of Aberdeen in Scotland had simultaneously but independently of Davisson and Germer showed that an electron beam transmitted through a thin foil leads to diffraction, in agreement with de Broglie's formula (Thomson and Reid, 1927). However, Thomson was not nominated and could therefore not be considered for an award. In his special report, Oseen wrote that if the wave nature of electrons were to be recognized, it should be an undivided prize for Louis de Broglie. The Nobel Committee agreed, and on November 12, 1929, the Royal Swedish Academy of Sciences decided to award Louis de Broglie the 1929 Nobel Prize in Physics "för upptäckten av elektronernas vågnatur" ("for the discovery of the wave nature of electrons"). (One can note in passing that the reserved 1928 prize on the same day was awarded to Owen Richardson.)

The presentation speech on December 10, 1929, was given by Oseen in Swedish (see *Les Prix Nobel* 1929). It is of historical interest and is given in English translation at *nobelprize.org*.

Finally, an anecdote based on *Dagens Nyheter* of November 13, 1929, when the Nobel Prizes were reported: The phone call to Paris, to Louis de Broglie, first went by mistake to his older brother Maurice, also a physicist known for his work on X-ray spectroscopy. Maurice was happy to learn that his work on X-ray spectroscopy had been awarded the Nobel Prize, only to discover that it was for his brother's work on electrons that the award was given. When the mistake was sorted out, the phone call came through to Louis de Broglie, who first thought it was a mistake and that his brother Maurice had been awarded!

3.2 The Nobel Prize in 1937

As an epilogue to this chapter, it is worth noting that nominations for the experimental discovery of the wave nature of electrons continued in the 1930s. Thomson was first nominated in 1931, and in 1937 he received four nominations. Davisson received no less than a total of 44 nominations,

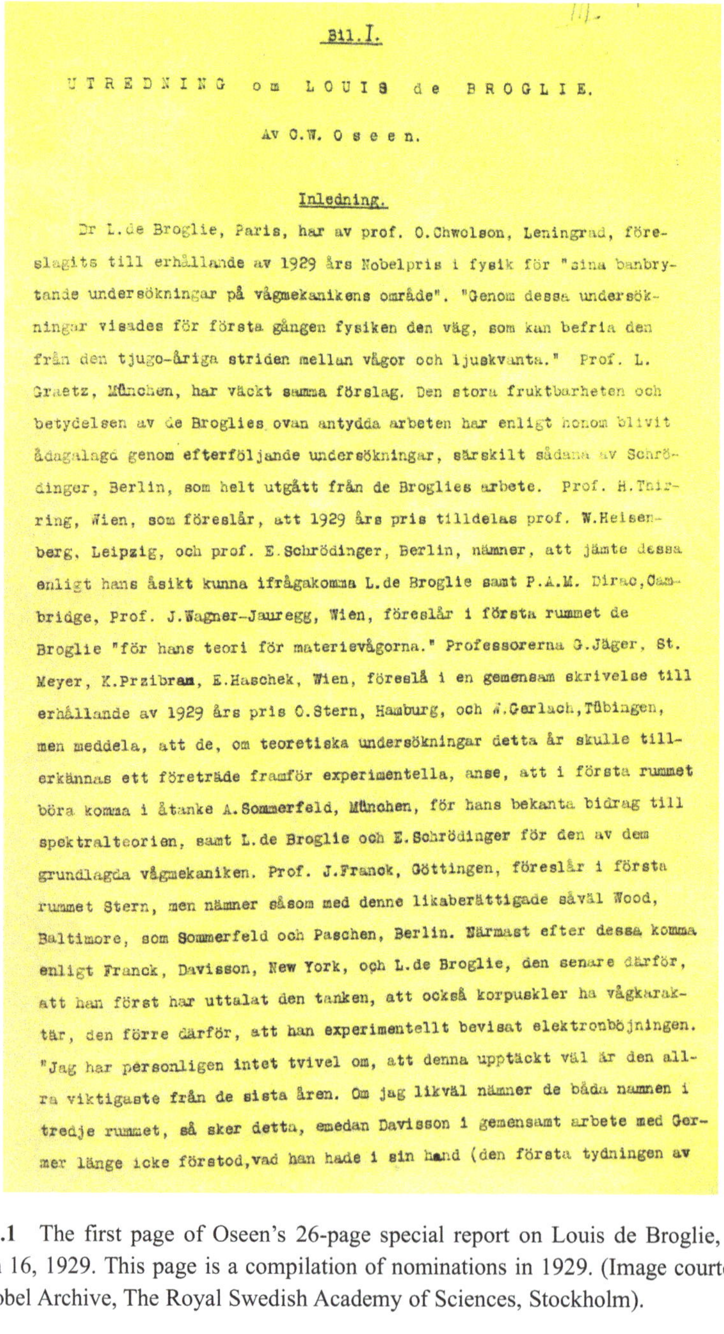

Bil. I.

U T R E D N I N G om L O U I S de B R O G L I E.

Av O.W. O s e e n.

Inledning.

Dr L.de Broglie, Paris, har av prof. O.Chwolson, Leningrad, föreslagits till erhållande av 1929 års Nobelpris i fysik för "sina banbrytande undersökningar på vågmekanikens område". "Genom dessa undersökningar visades för första gången fysiken den väg, som kan befria den från den tjugo-åriga striden mellan vågor och ljuskvanta." Prof. L. Graetz, München, har väckt samma förslag. Den stora fruktbarheten och betydelsen av de Broglies ovan antydda arbeten har enligt honom blivit ådagalagd genom efterföljande undersökningar, särskilt sådana av Schrödinger, Berlin, som helt utgått från de Broglies arbete. Prof. H.Thirring, Wien, som föreslår, att 1929 års pris tilldelas prof. W.Heisenberg, Leipzig, och prof. E.Schrödinger, Berlin, nämner, att jämte dessa enligt hans åsikt kunna ifrågakomma L.de Broglie samt P.A.M. Dirac, Cambridge, Prof. J.Wagner-Jauregg, Wien, föreslår i första rummet de Broglie "för hans teori för materievågorna." Professorerna G.Jäger, St. Meyer, K.Przibram, E.Haschek, Wien, föreslå i en gemensam skrivelse till erhållande av 1929 års pris O.Stern, Hamburg, och W.Gerlach, Tübingen, men meddela, att de, om teoretiska undersökningar detta år skulle tillerkännas ett företräde framför experimentella, anse, att i första rummet böra komma i åtanke A.Sommerfeld, München, för hans bekanta bidrag till spektralteorien, samt L.de Broglie och E.Schrödinger för den av dem grundlagda vågmekaniken. Prof. J.Franck, Göttingen, föreslår i första rummet Stern, men nämner såsom med denne likaberättigade såväl Wood, Baltimore, som Sommerfeld och Paschen, Berlin. Närmast efter dessa komma enligt Franck, Davisson, New York, och L.de Broglie, den senare därför, att han först har uttalat den tanken, att också korpuskler ha vågkaraktär, den förre därför, att han experimentellt bevisat elektronböjningen. "Jag har personligen intet tvivel om, att denna upptäckt väl är den allra viktigaste från de sista åren. Om jag likväl nämner de båda namnen i tredje rummet, så sker detta, emedan Davisson i gemensamt arbete med Germer länge icke förstod, vad han hade i sin hand (den första tydningen av

Fig. 3.1 The first page of Oseen's 26-page special report on Louis de Broglie, dated March 16, 1929. This page is a compilation of nominations in 1929. (Image courtesy of the Nobel Archive, The Royal Swedish Academy of Sciences, Stockholm).

including 9 in 1937. Thomson and Davisson shared the 1937 Nobel Prize "for their experimental discovery of the diffraction of electrons by crystals."

Germer, who received a total of 26 nominations, including 8 in 1937, was not selected for the award. By 1937, it was clear to the Nobel Committee that Thomson's work on the application of electron diffraction had become important and that electrons had now started to compete with X-rays. The question as to why Germer was excluded does not have an answer. In the report from the Nobel Committee to the Academy, Germer is mentioned several times together with Davisson, but the report does not provide an explicit reason for his exclusion.

1929
de Broglie, Louis-Victor
1892 -
France
"För upptäckten av elektronernas vågnatur".

Chapter 4

THE QUANTUM MECHANICS NOBEL PRIZES IN 1932/33

4.1 Introduction

On November 13, 1930, Chandrasekhara Venkata Raman from India became the first laureate outside Europe and the US to be awarded the Nobel Prize. The next announcement of Nobel Prizes in Physics did not come until November 9, 1933, when it was revealed that Werner Heisenberg had been awarded the Nobel Prize reserved for 1932 and that the 1933 Nobel Prize was shared between Erwin Schrödinger and Paul Dirac. This remains the longest peacetime gap in the history of the Physics Prize.

It came as a total surprise to me when I began looking into the history of the prizes awarded to the three individuals who arguably made the most important contributions to quantum mechanics — and who significantly enhanced the prestige of the Nobel Prize in Physics. In this chapter, I will try to shed some light on this three-year desert trek.

4.2 The 1930 Nobel Prize

The Nobel Prize awarded to Raman in 1930 is the only one that stands between the first recognition of the new quantum mechanics — given to Louis de Broglie in 1929 — and the award in 1933. Raman's discovery of the Raman effect in 1928 came as a complete surprise, but was quickly recognized as something very useful.

In his Nobel lecture, delivered in Stockholm on December 11, 1930 (*Les Prix Nobel* 1930), and held in his native scientific language, English, Raman told the story of how he first became fascinated by light scattering during a boat trip in the Mediterranean in 1921. While watching the sunlight scatter on the water, he became intrigued. Back at Calcutta University, he and his collaborators began exploring a wide range of light-scattering phenomena.

The two breakthrough papers appeared in 1928, and their significance was immediately clear to the physics community. Raman was nominated for the Nobel Prize in 1929 by Niels Bohr, and in 1930 he received ten nominations — six of them from Nobel laureates, including Bohr himself. His surprising discovery was a perfect fit for a Nobel Prize and presented no difficulties for the Nobel Committee.

Although the quantum-mechanical aspects of Raman's discovery were noted in the December 10 presentation speech by Henning Pleijel, the focus was placed on its practical value — the use of the Raman effect to study molecular structures. When the primary light beam is scattered, it produces spectral lines at slightly shorter or longer wavelengths than the original beam. This meant that instead of struggling to detect molecular radiation directly in the near-infrared region — where photographic plates of the time were nearly useless — scientists could observe the secondary Raman-scattered light in a region where the photographic plates were far more sensitive.

The importance of the Raman effect for chemistry was recognized by *Dagens Nyheter* immediately after the announcement, as shown in Fig. 4.1. Nearly a century later, this fact remains just as true.

It follows from the abstract of the article in *Nature* (vol. 121, pp. 501–502, 1928) by Raman and K. S. Krishnan that they were inspired by Compton's discovery of the effect that now bears his name. Their article in *Nature* appeared on March 31, 1928, less than four months after Compton had been awarded the Nobel Prize. In it, they argued that just as X-rays scatter inelastically on electrons and lose energy, so too a transfer of energy should occur when ordinary light is scattered by atoms or molecules — and they were right.

Only a couple of weeks earlier, on March 16, 1928, Raman had given his Inaugural Address to the South Indian Science Association, later published in the *Indian Journal of Physics* (vol. 2, pp. 387–398, 1928). At the end of this talk, he speculated that the effect should also occur for X-rays. This prediction, too, turned out to be correct.

DAGENS NYHETER.

Fredagen den 14 November 1930

Två Nobelpristagare

"Vägen tillbaka"

Remarques nya bok, läses som följetong i Dagens Nyheter.

Fortsättning på frontskildringen.

Författaren uttalar sig.

Krigsgenerationen i fredens värld bokens problem.

En indisk fysiker och en tysk kemist prisbelönade i år.

Kaféstriden vid Nygatan har bilagts.

Ramans arbete även av vikt för kemin.

Lika skatt föreslagen, 5 kr. 25 öre.

D:r Paulsson tar sin hand från London

Enigheten

Två dödsoffer för explosion i Avestaverk.

Föga hopp finnes

Fig. 4.1 The front page of the daily national Swedish newspaper *Dagens Nyheter* on November 14, 1930. In English: "Two Nobel laureates. An Indian physicist and a German chemist awarded this year. Raman's work of importance also to chemistry." The German chemist is Hans Fisher. Used with permission.

The vast literature on Raman spectroscopy is, of course, far beyond the scope of this book. For a concise introduction, Wikipedia articles on Raman spectroscopy and X-ray Raman scattering provide a useful starting point.

4.3 Heisenberg's Paper on the New Quantum Mechanics — the Very First

The story of how Werner Heisenberg developed the new quantum theory is well known. In the summer of 1925, shortly after becoming a

Privatdozent in Göttingen, he suffered from a severe attack of hay fever and asked Max Born, professor of theoretical physics, for a short vacation. He chose to spend it on the island of Helgoland, located about 70 kilometers off the German coast, where he hoped the fresh sea air would cure his fever. The isolation also gave him the chance to work more effectively than in Göttingen (Heisenberg, 1993). (For a demythologizing account of this story, see Dahn (2025).)

At the time, Heisenberg was determined to create a new theory that would replace the Bohr–Sommerfeld model. He realized that this meant abandoning the idea of electron orbits altogether. Since the position and period of an electron's revolution are unobservable, Heisenberg instead focused on quantities that could be observed — such as the frequencies (wavelengths) and intensities of spectral lines, or transition amplitudes. Back in Göttingen, he wrote down his findings in the paper "Über quantentheoretische Umdeutung kinematischer und mechanischer Beziehungen" ("Quantum-theoretical reinterpretation of kinematic and mechanical relations"), often referred to as the *Umdeutung* paper. It was received by *Zeitschrift für Physik* on July 29, 1925, and published in September of the same year (Heisenberg, 1925). The paper is notoriously difficult to follow, but the commentary by Aitchison, MacManus and Snyder (2004) is a considerable help. A broader perspective on the development of matrix mechanics can be found in MacKinnon (1977).

4.4 The Papers by Born, Heisenberg and Jordan

Having read a preprint of Heisenberg's paper, Max Born realized that the new quantum mechanics could be given a solid mathematical foundation through the use of matrices. Together with his student Pascual Jordan (Born and Jordan, 1925), he provided a mathematical framework for Heisenberg's relations between observables, expressed in terms of matrices. Heisenberg himself was not familiar with matrix algebra, but Born had learned it earlier from the mathematician Jakob Rosanes.

A follow-up paper was written by Born, Heisenberg and Jordan in 1926, known as the *Dreimännerarbeit* ("three-man paper"). Just like Heisenberg's first article, the Born and Jordan (1925) paper has continued to attract scholarly attention, with discussions as recent as Fedak and Prentis (2009).

Arguably, the most important equation in quantum mechanics — the canonical commutation relation — was derived in the *Dreimännerarbeit.* If q denotes the position operator and p the momentum operator, the canonical commutation relation is expressed as (Griffiths 2005, p. 43)

$$pq - qp = h/2\pi i$$

This equation is engraved on Max Born's gravestone in Göttingen. Heisenberg himself, however, was never entirely comfortable with the relation between quantum-mechanical quantities that did not commute.

The contributions of Born and Jordan to the new quantum mechanics — also known as matrix mechanics — have led to discussions about the Nobel Prize and, in particular, why they were left out of the 1932/33 awards. Born did eventually receive the Nobel Prize, but not until 1954 (as will be discussed in Chapter 5). I will return to this point later.

4.5 Dirac's First Paper on Quantum Mechanics

Paul Adrien Maurice Dirac studied at Bristol University in England from 1918 to 1923, first in electrical engineering and then in applied mathematics. He later moved to Cambridge, where he worked under Ralph Fowler as his supervisor. In July 1925, Werner Heisenberg visited Cambridge and gave a lecture on his new quantum theory. Fowler received a preprint of Heisenberg's paper, which he showed to Dirac. It took Dirac a remarkably short time to formulate his own version of quantum mechanics, and his paper (Dirac, 1925) was submitted to *Proceedings of the Royal Society* on November 7, 1925.

Michael Berry (1998), renowned particularly for his work on the geometric phase, wrote a beautiful feature article on Dirac in *Physics World.* In contrast to Heisenberg, Dirac had no difficulty with non-commuting objects — quantities representing measurable variables that should be regarded as operations. An experiment, in this view, is an operation that produces a number. As Berry illustrates, there is nothing strange about the fact that putting on socks before shoes leads to a different result than putting on shoes before socks.

With Dirac's first paper and those that followed, quantum mechanics took the definitive form that is still in use today.

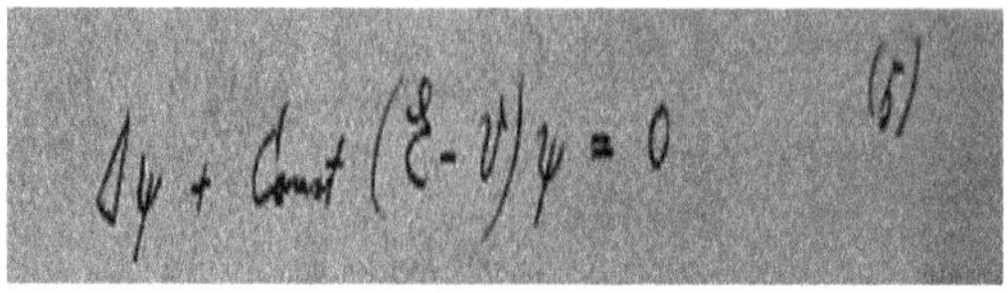

Fig. 4.2 The Schrödinger equation in Schrödinger's own handwriting. Schrödinger Archive, Central Library of Physics, University of Vienna.

4.6 Schrödinger's Papers on Wave Mechanics

Erwin Schrödinger was inspired by Louis de Broglie's hypothesis on the wave nature of the electron. He argued that if the hypothesis was correct, there ought to exist a second-order differential equation analogous to the well-known wave equation for electromagnetic radiation. Schrödinger's first paper (Schrödinger, 1926) was submitted to *Annalen der Physik* on January 27, 1926. Notably, this was one year before experiments began to confirm de Broglie's hypothesis.

The first article was soon followed by five more papers in *Annalen der Physik*. English translations of these works, along with several additional papers by Schrödinger, are available in Schrödinger (1928). The wave equation is now known as the Schrödinger equation and is often used as the starting point for teaching quantum mechanics to undergraduate students (Griffiths 2005). Figure 4.2 shows a historic document — the very first Schrödinger equation, preserved in Schrödinger's own handwriting.

4.7 Oseen's Special Report in 1927

Oseen's special report in 1927, mentioned in Section 3.1, dealt primarily with the importance of the Compton effect for theoretical physics. Arthur H. Compton shared that year's Nobel Prize with Charles T. R. Wilson, who was honored "for his method of making the paths of electrically charged particles visible by condensation of vapor." It may be noted in passing that there was only a very weak connection between the two half-prizes; a strong connection is not required by the by-laws.

What makes Oseen's special report noteworthy is that he devoted several pages to Heisenberg's matrix mechanics and Schrödinger's wave mechanics, even though neither Heisenberg nor Schrödinger had yet been nominated. He also observed that in England, Dirac — likewise not yet nominated — had been working in parallel with Born, Heisenberg and Jordan, using a different approach to systematizing the theory.

Oseen acknowledged that Schrödinger's solution for the hydrogen atom — yielding the same result as Bohr's theory — was a major success. However, with no nominations for Dirac, Heisenberg or Schrödinger, he made no attempt to suggest that a Nobel Prize be awarded.

4.8 The First Nominations of Heisenberg and Schrödinger

In 1928, Erwin Meyer, a specialist in acoustics, nominated Heisenberg and Schrödinger for a Nobel Prize to be shared between them. Meyer was invited to submit a nomination in his capacity as chairholder at his university. That same year, Schrödinger was also nominated together with Petrus J. Debye by David S. Jordan, a foreign member of the Royal Swedish Academy of Sciences, who therefore held a permanent right of nomination. Debye would later, in 1936, be awarded a Nobel Prize — but in chemistry, not in physics.

4.9 The Mysterious "Nomination" in 1928 by Einstein

In Abraham Pais's book (1982a, p. 515), a nomination by Einstein dated September 25, 1928, is translated into English. In this nomination, Einstein suggested the possibility of a prize shared between Heisenberg, Born and Jordan. The date is not unusual, as Einstein seems to have had the habit of sending his nominations without waiting for the official award announcement in November.

It is worth noting that Pais acknowledged the assistance of Bengt Nagel, who at the time (in the early 1980s) was secretary of the Nobel Committee for Physics and knew the archive in every detail. In Nancy Greenspan's biography of Max Born, *The End of the Certain World*, Einstein's proposal is also mentioned on page 193 (Greenspan, 2005).

The fact that Einstein nominated Born, while Born himself was overlooked when Heisenberg received the reserved 1932 Nobel Prize, has also been discussed by Bernstein (2005).

When I first read the 1928 general report by the Nobel Committee for Physics to the Royal Swedish Academy of Sciences, I was puzzled by the fact that Einstein's nomination was nowhere mentioned. A nomination by Einstein could hardly have gone unnoticed. Around 2019, when I looked into this, neither Abraham Pais nor Bengt Nagel was alive. I consulted the secretaries who had succeeded Nagel, but none of them had ever heard anything about Einstein's nomination.

The next step was, of course, to search for the original nomination in the archive. Yet another mystery arose: in the list of incoming nominations, Einstein's name was nowhere to be found.

Finally, I received help from the late Lars Brink, who had longstanding contacts at Caltech. He suggested that I contact Diana Kormos Buchwald, General Editor and Director of *The Einstein Papers Project.* She provided the explanation. Einstein had indeed written a draft nomination, but it was never typed or sent to the Nobel Committee. The second page of Einstein's handwritten draft nomination letter is shown in Fig. 4.3. The date and the translation from German into English are correct, but how Pais obtained the handwritten letter — and why he assumed it had been sent — remains unknown and will most likely continue to be a mystery.

As we shall see in Section 5, Max Born was finally awarded a Nobel Prize, but more than 20 years later.

4.10 Ivar Waller's Letters to Oseen

In 1928, Dirac derived his relativistic wave equation, arguably his greatest scientific achievement. The first of two papers was submitted to *Proceedings of the Royal Society* on January 2 and published on February 1 (Dirac, 1928a). A second paper followed a month later (Dirac, 1928b).

Ivar Waller, who would later succeed Oseen as professor at Uppsala University in 1934 when Oseen moved to Stockholm, and who also replaced him on the Nobel Committee for Physics after Oseen's death in 1944, maintained frequent contact with him. Unlike Oseen, who traveled very little for health reasons, Waller traveled abroad and reported back in letters.

In May 1928, Waller visited Niels Bohr's institute in Copenhagen and wrote two letters to Oseen, dated May 3 and May 13. In the first letter, he described how Dirac had visited the institute and lectured on his most

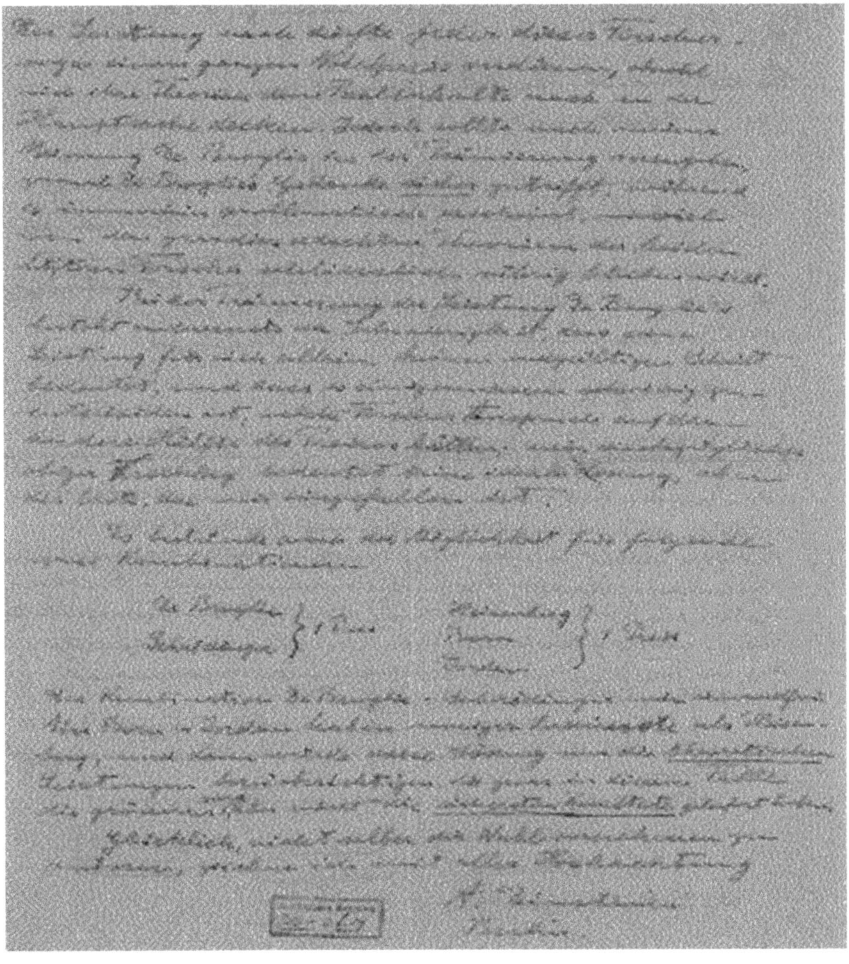

Fig. 4.3 Reproduced from *The Collected Papers of Albert Einstein*, Vol. 16, ed. Diana Kormos Buchwald (Princeton University Press, 2021). Used with permission.

recent work, the relativistic wave equation — the Dirac equation. Waller remarked that, in his opinion, Dirac had by then taken the leading role in quantum mechanics.

As Berry (1998) has pointed out, the Dirac equation for an electron moving in an arbitrary electromagnetic field was able to account for all the experiments known at the time, explain the origin of spin, and even predict the existence of antimatter. This last prediction caused the Nobel

Committee considerable concern over the next five years — since, at that time, no one had yet observed antimatter.

4.11 Oseen's Special Report, 1929

This special report, dated March 16, 1929, was discussed in Section 3.1 in connection with the Nobel Prize for de Broglie. Oseen, however, also devoted several pages to Heisenberg and Schrödinger. His conclusion was strikingly frank:

> It would not be an unreasonable demand, moreover, that a theory, in order to be considered worthy of a Nobel Prize, should have given rise to an experimental discovery of more fundamental nature. This condition is currently not met by either Heisenberg or Schrödinger.

It is worth noting that despite Waller's letter about Dirac, Dirac's publications of his relativistic wave equation (Dirac, 1928a, b) a year earlier, and his receiving his first nomination in 1929, Oseen made no mention of Dirac in this special report.

4.12 Svedberg's Nomination of Heisenberg in 1930

(Theodor) Svedberg, who was awarded the 1926 Nobel Prize in Chemistry, was for many years the most influential chemist in Sweden. The only reason to mention his nomination of Heisenberg here is that his nomination statement was preserved until Heisenberg was finally awarded the reserved 1932 Nobel Prize in Physics in 1933.

Svedberg's nomination emphasized Heisenberg's theoretical discovery of the allotropic forms of hydrogen (Heisenberg, 1927a). If I ask a physicist about the allotropic forms of hydrogen, they usually do not know what I am referring to. If, on the other hand, I ask about para- and ortho-hydrogen, they immediately recognize it (see Fig. 4.4). Heisenberg had discovered that the ratio of ortho- to para-hydrogen is 3:1 — these are, in the language of chemistry, the two allotropic forms of hydrogen.

Closely related to Heisenberg's discovery was David Dennison's calculation of the specific heat of hydrogen (Dennison 1927), which required him to assume that there were in fact two forms of hydrogen. Judging by the unassuming title of his paper, Dennison may not at the time have fully grasped the significance of his work for understanding proton spin.

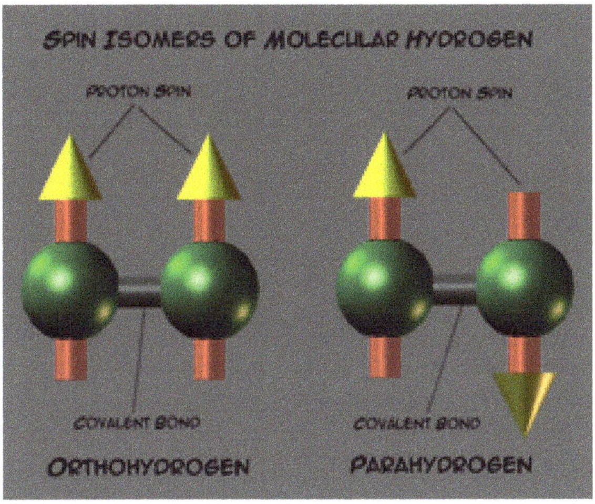

Fig. 4.4 In ortho-hydrogen the proton spins are in the same direction, whereas in para-hydrogen they are in opposite directions. At room temperature the ratio ortho/para is 3:1. In the language of chemistry, ortho and para are the two allotropic forms of hydrogen.

At the time of Heisenberg's and Dennison's papers, the stability of ortho- and para-hydrogen was not yet known — in other words, how rapidly the two forms could interconvert remained uncertain. The experimental breakthrough came two years later, when Bonhoeffer and Harteck (1929a, b) and Eucken (1929) succeeded in producing pure para-hydrogen, which proved to be stable on essentially any timescale.

Svedberg, who was well aware of Oseen's 1929 report (see Section 4.11) and, in particular, Oseen's view that neither Heisenberg's nor Schrödinger's theory had yet given rise to an experimental discovery of a more fundamental nature, now had everything he needed to respond. On January 31, 1930, he nominated Heisenberg — perhaps as a challenge to Oseen, or perhaps as support.

In his special report of 1930, Oseen concluded that the allotropic forms of molecular hydrogen belonged to chemistry rather than physics. Despite this, he devoted several pages to discussing hydrogen. He noted that if chemists considered the 1929 work of Bonhoeffer, Harteck and Eucken to be so important, they might consider awarding a Nobel Prize in Chemistry. The chemists, however, were of the view that a physics prize for Heisenberg should come before any chemistry award. Since Svedberg nominated both Bonhoeffer and Eucken (though not Harteck) for the chemistry prize in 1930, a chemistry award would at least have been possible in principle.

4.13 The Nobel Committee, 1930

The composition of the Nobel Committee changed quite drastically during the 1920s. At that time, a new member could be elected to the Royal Swedish Academy of Sciences only when an existing member passed away. Today, by contrast, a new member can be elected once they reach the age of 65. As a result, the pool of physicists available to serve on the Nobel Committee in the 1920s was very limited.

The members of the 1920 Nobel Committee were Svante Arrhenius (a member since the beginning of 1900), Bernard Hasselberg (since 1900), Gustaf Granqvist (since 1904), Vilhelm Carlheim-Gyllensköld (since 1910) and Allvar Gullstrand (since 1911). Hasselberg and Granqvist both passed away in 1922 and were replaced by Carl Wilhelm Oseen and Manne Siegbahn. The latter received the 1924 Nobel Prize in Physics (awarded in 1925) by a very slim margin — an interesting story, but outside the scope of this book.

Arrhenius passed away in 1927, while Gullstrand resigned from the Committee in 1929 and died the following year. Thus, by 1930, only Carlheim-Gyllensköld remained from the 1920 Committee. He was never a particularly influential member; in fact, he did not deliver a single presentation speech during the 1920s.

In 1928, Henning Pleijel, a specialist in theoretical electronics, became a member of the Committee, and two years later he was joined by Erik Hulthén, a molecular spectroscopist.

It is clear — and should already be evident from what has been discussed — that Oseen was the dominant member of the Nobel Committee.

4.14 The Nomination Situation in 1930 and Oseen's Special Report

By 1930, Heisenberg had received ten nominations, Schrödinger twenty-one, and Dirac just one. Schrödinger's wave mechanics was likely more accessible to most physicists than Heisenberg's matrix mechanics, although it was well understood that the two approaches yielded identical results. In present-day textbook language, these are referred to as the Schrödinger picture and the Heisenberg picture (Griffiths 2005, p. 130). Max Born — who would later receive the sixth Nobel Prize for quantum mechanics in 1954 — obtained his first nomination in 1930, while Wolfgang Pauli — who would go on to receive the fifth, in 1945 — had not yet been nominated.

Thus, there was considerable pressure from the scientific community, through nominations, to recognize quantum mechanics and its two leading figures. As described in Section 4.2, Raman's prize provided temporary relief for the Committee, but Oseen still had to continue his special reports on quantum mechanics. We have already seen how he avoided addressing Heisenberg's theoretical discovery of the statistical relation between ortho- and para-hydrogen.

In 1930, Oseen was also confronted with a nomination from Peter Pringsheim in Berlin for both Heisenberg and Born. He sidestepped Born's case by arguing that if Born were to be awarded, Jordan must also be included. However, since Jordan had not been nominated, the Heisenberg/Born combination was ruled out. Born's next nomination

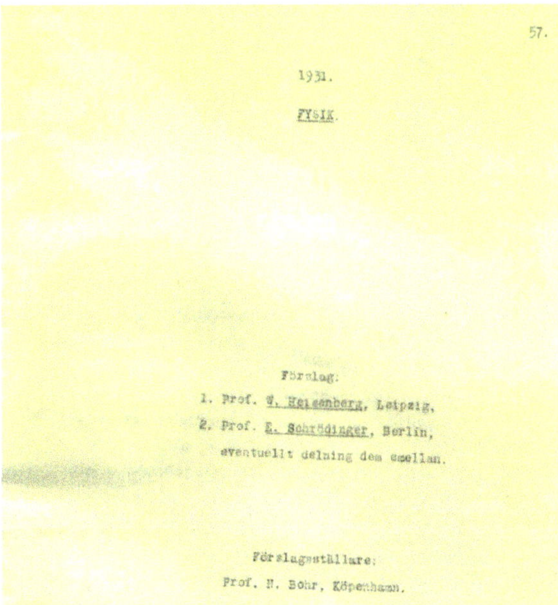

57.

1931.

FYSIK.

Förslag:

1. Prof. W. Heisenberg, Leipzig,
2. Prof. E. Schrödinger, Berlin,
eventuellt delning dem emellan.

Förslagsställare:

Prof. N. Bohr, Köpenhamn.

Fig. 4.5 Niels Bohr's nomination of Heisenberg and Schrödinger in 1931. Bohr uses only two sentences, the first one to argue that Heisenberg should be in the first position for an award and the second one to argue for Schrödinger's second position. (Image courtesy of the Nobel Archive, The Royal Swedish Academy of Sciences, Stockholm.)

58.

Inkom den 30. 1. 1931.

KØBENHAVN Ø 29. Januar 1931.

Til

Kungl. Vetenskapsakademiens Nobelkommitté för Fysik.

I Anledning af Komitéens Indbydelse til at indsende Forslag til Uddeling af Nobelprisen i Fysik 1931 skal jeg tillade mig at bringe Professor Werner Heisenberg i Forslag, som den der i første Linie har gjort sig fortjent ved Udviklingen af Kvanteteorien til en rationel Atommekanik, en Udvikling der har vaeret af afgørende Betydning for den teoretiske Udredning af de atomare Faenomener, hvorom Bestraebelserne indenfor vore Dages Experimentalfysik for saa stor en Del samler sig. Jeg skal dog i denne Forbindelse ikke undlade at minde om Professor Erwin Schrödingers store Fortjenester, der imidlertid efter mit Skøn maa komme i Betragtning i anden Linie, dersom ikke disse to fremragende Forskeres Fortjenester samtidig belønnes ved Nobelprisernes Uddeling.

Aerbødigst

Niels Bohr.

Fig. 4.5 (*Continued*)

came only in 1934, by which time the Nobel Prizes for quantum mechanics had already been awarded.

4.15 Niels Bohr's Nomination in 1931

When Niels Bohr finally turned his attention to the new quantum mechanics, he could look back on an impeccable nomination record during the 1920s: Albert Einstein (nominated by Bohr in 1920, awarded in 1922), James Franck (nominated in 1924–26, awarded in 1926), Owen Richardson (nominated in 1926 and 1928–29, awarded in 1929) and Chandrasekhara Raman (nominated in 1930, awarded in 1930). Since 1922, when Bohr himself received the Nobel Prize, he held permanent nomination rights. Bohr wrote his nominations in his native Danish, knowing that the Swedes would have no difficulty reading them. In his 1930 nomination, he placed Heisenberg in the first position and Schrödinger in the second.

4.16 Einstein's Nomination in 1931

Albert Einstein did not nominate as frequently as Niels Bohr during the 1920s, but his nomination record was impeccable as he entered the 1930s. He nominated Max Planck in 1919 (awarded the same year), James Franck and Gustav Hertz in 1924 (awarded in 1926) and Arthur Compton in 1926 (awarded in 1927). As noted in Section 4.9, his 1928 nomination was only a draft and was never sent to the Nobel Committee.

If one consults Einstein's nominations at *nobelprize.org*, there is no entry for 1931. The reason is that Einstein, as mentioned earlier, sometimes submitted his nominations before the outcome of the preceding year was known. In those days, the decision by the Royal Swedish Academy of Sciences was made in November, while invitation letters to nominate for the following year were sent in September. Beginning in 1931, the Nobel Committee's report to the Academy was dated August 31, whereas Einstein's nomination letter that year arrived in Stockholm on October 10. As a result, it was listed as a nomination for 1932.

Einstein's nomination letter was carefully written in German and is shown in Figure 4.6. The English translation is taken from Pais (1982, p. 515):

> In my opinion, this theory [quantum mechanics] contains without doubt a piece of the ultimate truth. The achievements of both men are independent of each other and so significant that it would not be appropriate to divide a Nobel prize between them. The question of who should get the prize first is hard to answer. Personally, I assess Schrödinger's achievement as the greater one, since I have the impression that the concepts created by him will carry further than those of Heisenberg. [Here Einstein adds a footnote: This, however, is only my opinion, which may be wrong.] On the other hand, the first important publication by Heisenberg precedes the one by Schrödinger. If I had to decide, I would give the prize first to Schrödinger.

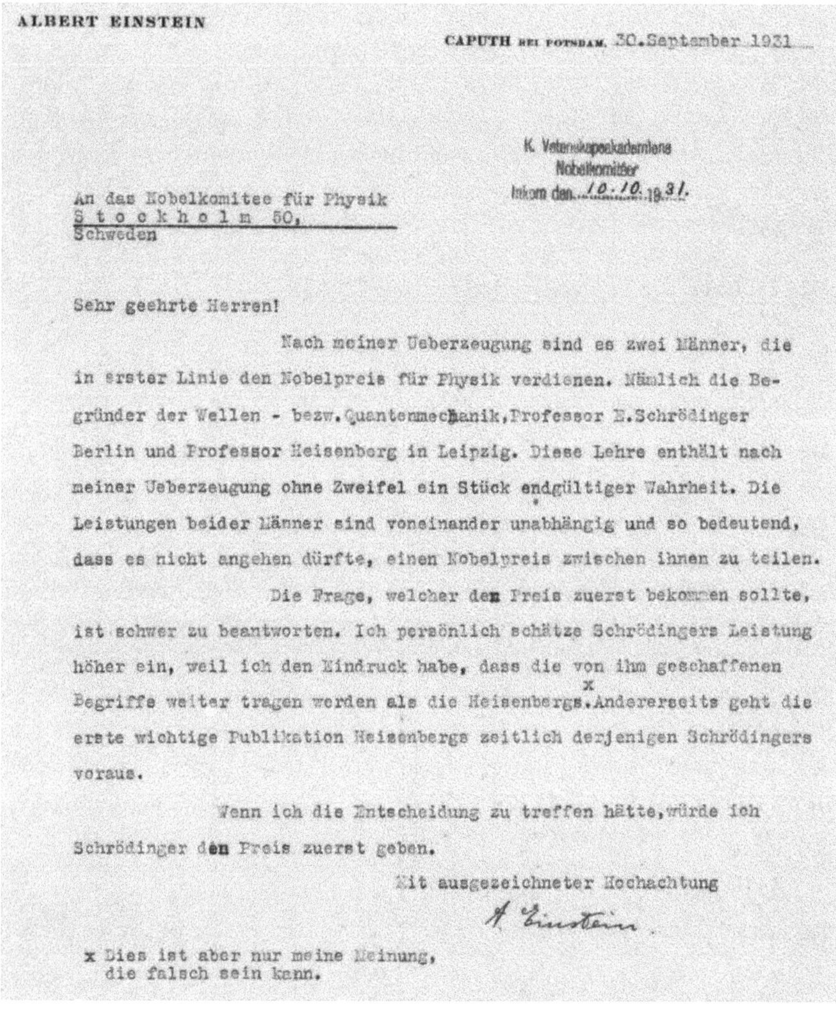

ALBERT EINSTEIN

CAPUTH BEI POTSDAM, 30.September 1931

K. Vetenskapsakademiens
Nobelkommittér
Inkom den 10.10.1931.

An das Nobelkomitee für Physik
S t o c k h o l m 50,
Schweden

Sehr geehrte Herren!

Nach meiner Ueberzeugung sind es zwei Männer, die in erster Linie den Nobelpreis für Physik verdienen. Nämlich die Begründer der Wellen - bezw. Quantenmechanik, Professor E.Schrödinger Berlin und Professor Heisenberg in Leipzig. Diese Lehre enthält nach meiner Ueberzeugung ohne Zweifel ein Stück endgültiger Wahrheit. Die Leistungen beider Männer sind voneinander unabhängig und so bedeutend, dass es nicht angehen dürfte, einen Nobelpreis zwischen ihnen zu teilen.

Die Frage, welcher den Preis zuerst bekommen sollte, ist schwer zu beantworten. Ich persönlich schätze Schrödingers Leistung höher ein, weil ich den Eindruck habe, dass die von ihm geschaffenen Begriffe weiter tragen werden als die Heisenbergs.x Andererseits geht die erste wichtige Publikation Heisenbergs zeitlich derjenigen Schrödingers voraus.

Wenn ich die Entscheidung zu treffen hätte, würde ich Schrödinger den Preis zuerst geben.

Mit ausgezeichneter Hochachtung

A. Einstein

x Dies ist aber nur meine Meinung, die falsch sein kann.

Fig. 4.6 Einstein's nomination letter dated September 30, 1931, and written in his house at Caputh bei Potsdam. The house is still there and well preserved. (Image courtesy of the Nobel Archive, The Royal Swedish Academy of Sciences, Stockholm.)

4.17 Oseen's Special Report 1931 and the Report of the Nobel Committee

Oseen wrote an extensive special report — 31 pages long — for the Nobel Committee, dated April 15, 1931. In it, he gave a detailed account of

Heisenberg's individual papers, the joint paper by Born and Jordan, and the collaborative paper by Heisenberg, Born and Jordan. He also discussed Schrödinger's contributions and concluded that Schrödinger's wave mechanics and the Heisenberg–Born–Jordan quantum (matrix) mechanics differed only in their mathematical frameworks.

What, then, was gained by Schrödinger's contribution? Oseen's answer was (translated from Oseen's original Swedish text):

> A new method for treating problems in atomic physics, a method that uses the same mathematical apparatus as the theory of differential equations, which in the older branches of physics had already been put to use. The knowledge of ordinary and partial differential equations that has been gained during 200 years of mathematical research was, through Schrödinger, put into the service of atomic physics.

Oseen discusses for the first time one of Heisenberg's most important contributions: his uncertainty principle (Heisenberg 1927b).

The title of Oseen's special report in Swedish is "Utredning rörande W. Heisenberg och E. Schrödinger." Dirac is mentioned only in passing at a few places: "At the same time as the studies by Born and Jordan, as well as by Born, Heisenberg, and Jordan, investigations into the same questions were being conducted by Dirac. I see no reason to provide an account of these Dirac works here."

Oseen argued that one of the most important papers in the new atomic theory — at least with respect to the problem it addressed — was the joint work by Heisenberg and Wolfgang Pauli (1929). In translation, Oseen wrote:

> Both in matrix mechanics and in wave mechanics, the energy expression has in most cases been derived directly from classical physics. A consequence of this has been that generally no consideration has been given to the fact that energy spreads at a finite speed. In this regard, bringing theory into better alignment with reality is the goal that both authors set for themselves. In order to achieve this, they must give the theory an invariant form with respect to Lorentz transformations. It may seem as if this would not happen without a solution to the difficulties revealed by Dirac's electron theory. However, the two authors believe that a solution to their problem is possible without addressing the deeper underlying difficulties.

Thus, Oseen was clearly aware of Dirac's relativistic wave equation of 1928, while also noting that difficulties remained with this theory — though he did not specify what these problems were. We will return to Dirac later; in 1931, his greatest triumph was still two years away.

The Nobel Committee was unanimous: the 1931 prize would be reserved for 1932. It is striking to note that the challenges of reconciling quantum theory with relativity were now used as an argument against awarding a prize to quantum theory and its leading figures, Heisenberg and Schrödinger. Ten years earlier, the Committee had likewise been unable to honor Einstein for what was his greatest achievement, the theories of relativity.

4.18 1932 and No Consensus in the Nobel Committee

In 1932, the Nobel Committee had two prizes at its disposal: the reserved prize from 1931 and the regular prize for 1932. Bohr took advantage of this situation by nominating Heisenberg for the 1931 prize and Schrödinger for the 1932 prize. By this time, Einstein's nomination was also on the table (Fig. 4.6). The number of nominations for both Heisenberg and Schrödinger increased compared with 1931. Heisenberg received seven nominations in 1932, bringing his total from 1928 to 19. Schrödinger received six, raising his total to 30. More significant than the raw numbers, however, was the stature of their nominators. Heisenberg had been nominated by Einstein, Bohr and Planck (in 1930). Schrödinger, too, had been nominated by Einstein, Bohr and Planck, and in addition by several other Nobel laureates: de Broglie (1931), Max von Laue (1914 laureate in Physics; nomination of Schrödinger in 1930) and Jean Perrin (1926 laureate in Physics; nomination in 1929). Swedish theoretical physicists Oskar Klein and David Enskog also nominated both Heisenberg and Schrödinger in 1932. By contrast, Dirac had received only a single nomination up to that point — by Hans Thirring in 1928 — and even then, Dirac was listed as a second choice after Heisenberg and Schrödinger. It is difficult to understand why Dirac attracted so little attention from the physics community at the time, but the reasoning behind individual nominations — meant to be strictly confidential — remains obscure.

In 1932, it was absolutely clear to the Nobel Committee that awarding anyone other than Heisenberg or Schrödinger was unthinkable. For the first time, the Committee commissioned a special report from someone other than Oseen — namely Erik Hulthén, a molecular spectroscopist.

He was asked to review quantum mechanics and experimental research. As an expert in molecular spectroscopy, Hulthén focused primarily on progress in his own field of interest. This did not convince the remaining four members of the Committee.

Oseen submitted a two-page statement in which he argued that neither Heisenberg nor Schrödinger had accomplished anything that could be classified as a "discovery or invention," as required by Alfred Nobel's will. In his view, not a single nomination referred to a genuine discovery, with the sole exception of Svedberg's 1930 nomination concerning the allotropic forms of hydrogen.

At the end of his statement, Oseen referred to the discovery of the neutron, which he suggested might prove important for overcoming some of the difficulties facing the new theory. How significant the discovery would turn out to be, he noted, remained to be seen. The neutron had been discovered by James Chadwick, whose results were published in June 1932 (Chadwick, 1932). Thus, Oseen was aware of the discovery when he wrote his two-page statement in September that year. Chadwick was awarded the Nobel Prize three years later for this achievement.

Hulthén also submitted a special statement advocating that the 1931 prize be divided between Heisenberg and Schrödinger and that the 1932 prize be reserved. Hulthén, however, ended up on the losing side — largely through his own shortcomings. His report failed to inspire. Oseen, dissatisfied with the unresolved problem of reconciling quantum mechanics with relativity, persuaded the Committee to postpone the 1932 award until 1933.

This decision meant that the 1931 prize was lost, since a Nobel Prize could be reserved for only one year in advance, not two. The prize money was transferred to a special fund — something that, one can assume, Manne Siegbahn found satisfactory (e.g., Friedman 2001).

4.19 1933: The Grand Finale

When the Nobel Committee entered 1933, it was clear that a prize for quantum mechanics could no longer be postponed. Bohr repeated his nominations from the previous year, recommending the prize reserved for 1932 for Heisenberg and the 1933 prize for Schrödinger. Einstein, by contrast, nominated only Schrödinger, who that year received 11 nominations — bringing his total from 1928 to 41. Heisenberg received 10 nominations in 1933, for a cumulative total of 29 since 1928.

For the first time since 1928, Dirac received not just one but two nominations — one from William Lawrence Bragg (Nobel Prize in Physics, 1915) and the other from the Russian-Polish physicist Czesław Białobrzeski. Both nominated Dirac alongside Heisenberg and Schrödinger. Bragg's handwritten nomination is shown in Fig. 4.7.

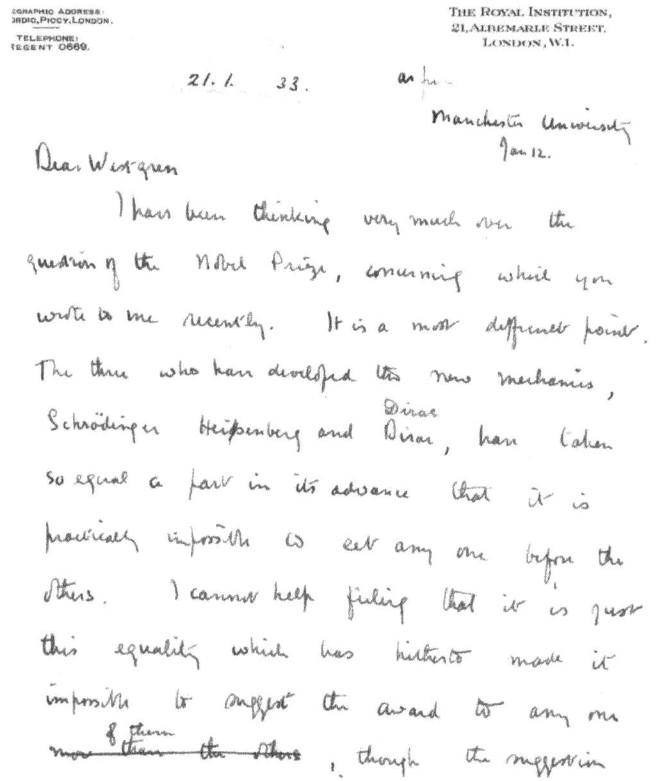

…GRAPHIC ADDRESS:
…RDIC, PICCY, LONDON.
TELEPHONE:
…EGENT 0669.

THE ROYAL INSTITUTION,
21, ALBEMARLE STREET,
LONDON, W.1.

21. 1. 33.

Manchester University
Jan 12.

Dear Westgren

I have been thinking very much over the question of the Nobel Prize, concerning which you wrote to me recently. It is a most difficult point. The three who have developed the new mechanics, Schrödinger Heisenberg and Dirac, have taken so equal a part in its advance that it is practically impossible to set any one before the others. I cannot help feeling that it is just this equality which has hitherto made it impossible to suggest the award to any one of them ~~more than the others~~, though the suggestion would otherwise have been made. In the

Fig. 4.7 William Lawrence Bragg's nomination of Heisenberg, Schrödinger and Dirac, addressed to Arne Westgren, the secretary of the Nobel Committee. Bragg cannot make any distinction between them, and suggests that one of the prizes at the Committee's disposal be divided between Heisenberg and Schrödinger, and Dirac be awarded half the other prize available to the committee. Bragg does not offer any suggestions as to who should have the other half. In his impressive biography of Dirac, Kragh (1990) argues that Bragg did not nominate Dirac as a first choice, but it is clear from the letter that Bragg makes no distinction between the three. It is also clear from the Nobel Committee's report in September 1933 that they correctly understood Bragg's suggestion. (Image courtesy of the Nobel Archive, The Royal Swedish Academy of Sciences, Stockholm.)

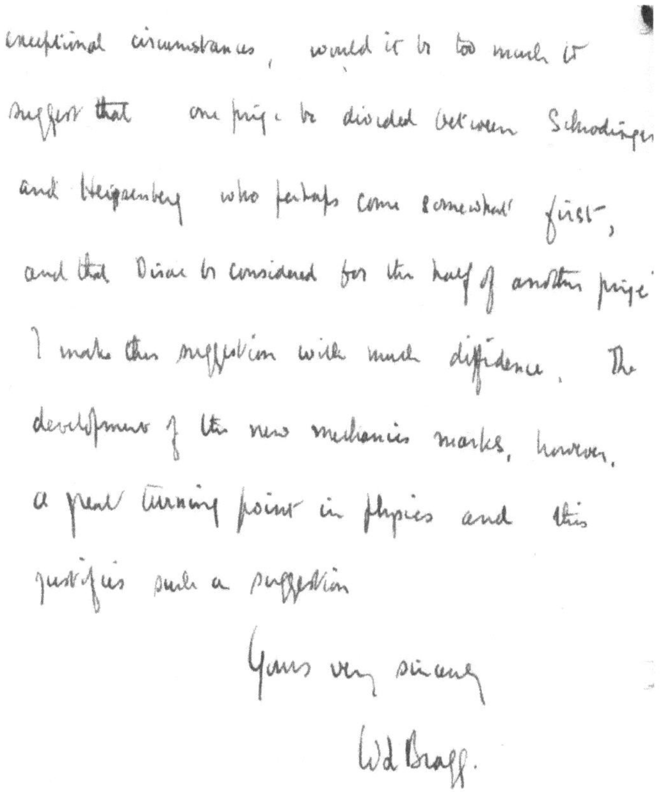
exceptional circumstances, would it be too much to suggest that one prize be divided between Schrodinger and Heisenberg who perhaps come somewhat first, and that Dirac be considered for the half of another prize? I make this suggestion with much diffidence. The development of the new mechanics marks, however, a great turning point in physics and this justifies such a suggestion

Yours very sincerely

W L Bragg.

Fig. 4.7 (*Continued*)

Oseen nominated Heisenberg and Schrödinger together with Wolfgang Pauli, who thus received his first nomination (many more would follow until he was awarded the Nobel Prize in 1945; see Chapter 5). By including Pauli, Oseen ensured that the Committee could formally consider him, since the required minimum of one nomination had now been met. It is less clear why Oseen also included Heisenberg and Schrödinger, both of whom already had several nominations in 1933. Bragg's nomination must have influenced the Committee, as is evident from its report to the Academy in September 1933, where Bragg's nomination is explicitly mentioned, while that of Białobrzeski is not. Oseen, who had previously referred to Dirac only in passing in his special reports, now prepared a detailed 28-page report on Dirac's scientific work.

Before turning to the details of Oseen's special report, it is worth noting that 1932 was an extraordinary year for major discoveries. The discovery of the neutron has already been mentioned, but it was also the year Carl Anderson discovered the positron (positive electron; Anderson, 1932). Anderson's 1932 paper, however, was brief and ambiguously titled, and presented only tentative interpretations. Oseen and the Committee can hardly be faulted for not recognizing its significance at the time. As we shall see shortly, Anderson returned with a longer and clearer paper and was awarded half of the 1936 Nobel Prize for the discovery of the positron, the other half going to Victor Hess for his discovery of cosmic radiation. Another landmark of 1932 was Harold Urey's discovery of hydrogen with atomic weight 2, the so-called heavy hydrogen (deuterium), for which he received the Nobel Prize in Chemistry in 1934. That same year, John Cockcroft and Ernest Walton completed their accelerator and began their pioneering work on the transmutation of atomic nuclei by bombardment with artificially accelerated particles — work that eventually earned them the Nobel Prize in 1951.

These discoveries were not unrelated; for example, while the spin and magnetic moment of the neutron were difficult to measure directly, they could be determined indirectly through studies of the deuterium nucleus.

Oseen's special report on Dirac bore all the hallmarks of his style: long, detailed and critical. He compared Dirac's early work of 1925–26 on "English" quantum mechanics with the German formulation of Heisenberg, Born and Jordan. Oseen concluded that Dirac's version led to simpler and more elegant solutions than the more complicated German approach.

Oseen briefly discussed Dirac's important paper "The Quantum Theory of Emission and Absorption of Radiation" (1927). It is surprising that he failed to grasp the significance of this work, even though Dirac himself emphasized at the outset, "On the other hand, hardly anything has been done up to the present on quantum electrodynamics." It seems likely that Oseen's inability to recognize that Dirac had, almost single-handedly, laid the foundations of quantum electrodynamics was connected with his generally critical assessment of Dirac's contributions.

Oseen discussed at length Dirac's relativistic wave equation (Dirac, 1928a,b), which he assessed as the papers that more than any others had contributed to Dirac's fame. He readily acknowledged that the Dirac equation provided a correct description of electron spin. What he found more difficult to accept was its prediction of a particle with a positive charge and the same mass as the electron.

Oseen's "Summary and final reflections" reads as follows in translation:

> "When one surveys Dirac's work, the first impression is undoubtedly that it represents a remarkable labour-output. If one seeks to determine more closely the character of that work, one observes that a substantial part of it is devoted to carrying out Heisenberg's fundamental idea, and that this portion proceeds in parallel with work that was being done at the same time in Germany. By its very nature, the truly new element that Dirac was able to contribute on this side of his activity lies mainly in method — in the mathematical-physical technique — something that, of course, does not preclude his occasionally surpassing his German colleagues in the results achieved. Achievements that bear the stamp of novelty in a higher sense were possible for Dirac only when the phase just mentioned, dominated by Heisenberg, was drawing to a close. Such achievements are: the theory of the one-electron problem (1928), the theory of magnetic poles (1931), and the attempt at a new foundation of quantum mechanics (1932). To date only the first of these theories has acquired real significance for physics, and it has occupied contemporary researchers to a high degree. In spite of all that has been written about this theory, it must still, on the whole, be regarded as science's latest word on the question — yet not its definitive word. That can be uttered only after clarity has been gained about the basis of the difference between positive and negative electricity. If one asks whether Dirac is a scientific innovator of the same stature as a Planck, an Einstein, or a Bohr, my answer for the present would be an unequivocal "no." But one must remember that becoming a great innovator depends not only on the researcher himself; it also depends on the era in which he lives. When Dirac first opened his eyes to the scientific world, the most important visible task before him was undoubtedly the elaboration of Heisenberg's idea. Dirac devoted himself to this task with heart and soul. Thus far, it has not left him time for innovative work on a grand scale. Yet it is by no means impossible that such work may still emerge. It is noteworthy that Dirac's most original contributions stem from the last few years."
>
> Uppsala in March 1933
> C.W. Oseen

It is clear that Bragg's nomination of Dirac, placing him on equal footing with Heisenberg and Schrödinger, required the preparation of a special report on Dirac. It is equally clear that Oseen's March 1933 report made it evident that Dirac was not considered a serious contender for a Nobel Prize that year. Oseen's assessment can be contrasted with a much later evaluation by the great theoretical physicist and 1965 Nobel laureate Sin-Itiro Tomonaga (1997, p. 59):

> "We mortals are left reeling by this staggering outpouring of ideas from Dirac. His first requirement is that the [Dirac] equation be linear in time. Such thoughts had not occurred to Schrödinger, Klein, Gordon, or even Pauli."

Tomonaga's admiration for Dirac's genius reflects the view that most physicists today would readily share.

The discovery of the positron

The interval between Oseen's special report on Dirac and the remarkable development of the positron — which eventually forced Oseen to change his mind — deserves a section of its own.

I have already mentioned Anderson's brief discovery paper of 1932. A remarkable coincidence occurred in March 1933, the same month that Oseen delivered his special report. This timing suggests that Oseen was unaware of Anderson's short communication in *Science* — or, alternatively, that he was neither impressed nor convinced by it.

On March 15, 1933, Anderson published his longer and definitive paper in *Physical Review* under the title "The Positive Electron." Figure 1 of this paper has since acquired iconic status: it shows a Wilson cloud chamber track, recorded under a strong magnetic field, in which an electron and a positron produced by cosmic rays curve in opposite directions — leaving no doubt that the positron had been discovered.

There was no immediate reaction to Anderson's discovery. When the Nobel Committee met in June 1933, Oseen had prepared a short report titled "Om den nya atomfysiken" ("Concerning the New Atomic Physics"); see Figure 4.8. In it, only Heisenberg and Schrödinger were mentioned, and it is clear that Oseen was now prepared to award them the prizes for 1932 and 1933. It is equally clear that Anderson's discovery had not yet made any impression on him.

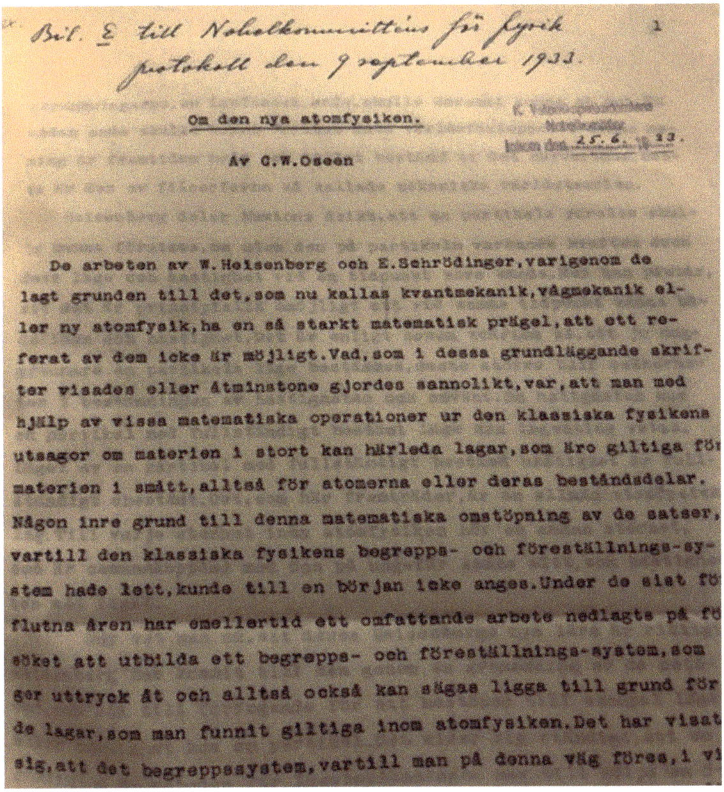

Bil. B till Nobelkommitténs för fysik protokoll den 9 september 1933. 1

Om den nya atomfysiken.

K. Vetenskapsakademiens Nobelkommittéer Inkom den 25.6. 19 33.

Av C.W.Oseen

De arbeten av W.Heisenberg och E.Schrödinger,varigenom de lagt grunden till det,som nu kallas kvantmekanik,vågmekanik eller ny atomfysik,ha en så starkt matematisk prägel,att ett referat av dem icke är möjligt.Vad,som i dessa grundläggande skrifter visades eller åtminstone gjordes sannolikt,var,att man med hjälp av vissa matematiska operationer ur den klassiska fysikens utsagor om materien i stort kan härleda lagar,som äro giltiga för materien i smått,alltså för atomerna eller deras beståndsdelar. Någon inre grund till denna matematiska omstöpning av de satser, vartill den klassiska fysikens begrepps- och föreställnings-system hade lett,kunde till en början icke anges.Under de sist fö flutna åren har emellertid ett omfattande arbete nedlagts på fö söket att utbilda ett begrepps- och föreställnings-system,som ger uttryck åt och alltså också kan sägas ligga till grund för de lagar,som man funnit giltiga inom atomfysiken.Det har visat sig,att det begreppssystem,vartill man på denna väg föres,i vi

Fig. 4.8 Oseen's report "Om den nya atomfysiken" ("Concerning the New Atomic Physics"). The handwritten statement at the top — "Appendix B to the protocol of the Nobel Committee for Physics, September 9, 1933" (in translation) — may seem confusing, but will be explained in the text and Fig. 4.9. (Image courtesy of the Nobel Archive, The Royal Swedish Academy of Sciences, Stockholm.)

The physics community, however, reacted quickly. In 1933, Robert Oppenheimer and Milton Plesset proposed that pairs of negative and positive electrons could be created in the vicinity of a nucleus irradiated by gamma rays. Their theory provided quantitative verification of the observations.

At this point, Oseen had everything he could possibly ask for: the unification of special relativity and quantum mechanics, together with a completely unexpected prediction that had been confirmed by experiment. When the Nobel Committee met again in September, Dirac's name was added by hand to Oseen's report from June (see Fig. 4.9).

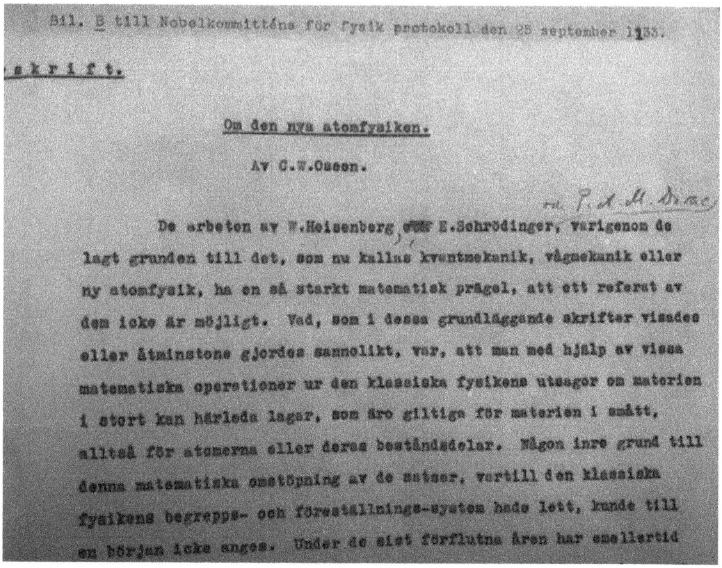

Bil. B till Nobelkommitténs för fysik protokoll den 25 september 1933.

skrift.

Om den nya atomfysiken.

Av C.W.Oseen.

och P.A.M. Dirac

De arbeten av W.Heisenberg, och E.Schrödinger, varigenom de lagt grunden till det, som nu kallas kvantmekanik, vågmekanik eller ny atomfysik, ha en så starkt matematisk prägel, att ett referat av dem icke är möjligt. Vad, som i dessa grundläggande skrifter visades eller åtminstone gjordes sannolikt, var, att man med hjälp av vissa matematiska operationer ur den klassiska fysikens utsagor om materien i stort kan härleda lagar, som äro giltiga för materien i smått, alltså för atomerna eller deras beståndsdelar. Någon inre grund till denna matematiska omstöpning av de satser, vartill den klassiska fysikens begrepps- och föreställnings-system hade lett, kunde till en början icke anges. Under de sist förflutna åren har emellertid

Fig. 4.9 When the Nobel Committee met again on September 25, 1933, Dirac had finally met the high standard required by Oseen, and Dirac's name was added to Oseen's report by hand. (Image courtesy of the Nobel Archive, The Royal Swedish Academy of Sciences, Stockholm.)

Friedman claims in his book (2001, p. 175) that it was Ivar Waller who set Oseen on the right track, but I have been unable to find any proof of this. Waller's two letters to Oseen were mentioned earlier and date back to 1928. No doubt, the Nobel Committee and Oseen followed what happened during the summer of 1933.

The Nobel Committee was unanimous when it met on September 25, 1933, deciding to award the prize reserved for 1932 to Heisenberg alone and to divide the 1933 prize between Schrödinger and Dirac. Figure 4.10 shows the final page of the report submitted to the Academy, where handwritten amendments reveal last-minute changes. In Heisenberg's case, the reference to the allotropic forms of hydrogen survived in the citation. At that time, only the citation in Swedish was given: "*För uppställande av kvantmekaniken, vars användning bland annat lett till upptäckten av vätets allotropa former.*" This is the hand-modified version from the report. In the original typed draft, the phrase "*särskilt*" ("in particular") was used, but the Committee evidently felt this was too strong and changed it to "*bland annat*" ("inter alia"). An English translation was

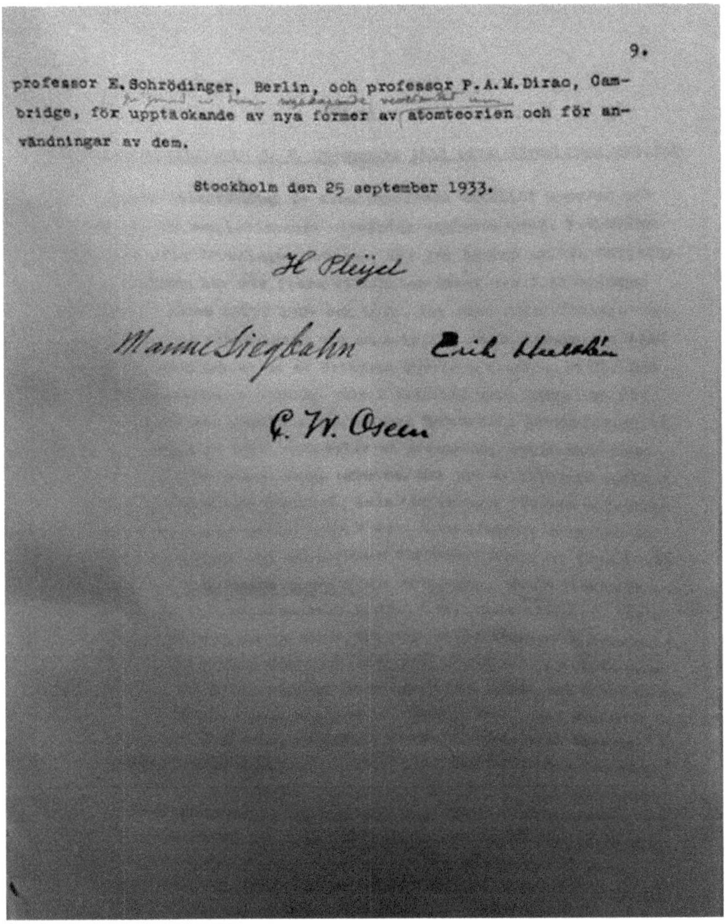

9.

professor E. Schrödinger, Berlin, och professor P.A.M. Dirac, Cambridge, för upptäckande av nya former av atomteorien och för användningar av dem.

Stockholm den 25 september 1933.

H Pleijel

Manne Siegbahn Erik Hulthén

C. W. Oseen

Fig. 4.10 The last page of the report prepared by the Nobel Committee and submitted to the Royal Swedish Academy of Sciences. The Physics Class supported the proposal, and on November 9, 1933, the academy made the decision in accordance with the committee's report. Almost three years had passed since Raman was awarded the 1930 Nobel Prize in Physics, the longest peacetime gap in the history of the physics prize. (Image courtesy of the Nobel Archive, The Royal Swedish Academy of Sciences, Stockholm.)

produced much later. It reads: "for the creation of quantum mechanics, the application of which has, inter alia, led to the discovery of the allotropic forms of hydrogen." Personally, I would prefer "for the formulation of quantum mechanics" rather than "for the creation," but the official

KAROTTER
HALLBERG

SVENSKA DAGBLADET

NY AMERIKANSK CIGARETT
"MARVELS" 6½ öre
HUGO LUNDGREN, Tobaksimport

Luthers släkt lever i Sverige

med över 1200 representanter på spinnsidan.

LUTHERS VIGSELRING BEVARAD.

Fyra Nobelpristagare utsedda

Fysikprisen
till
W. HEISENBERG, Leipzig.
E. SCHRÖDINGER, Berlin.
P. A. M. DIRAC, Cambridge.

Litteraturpriset
till
IVAN BUNIN.

Ivan Bunin hämtar själv sitt Nobelpris.

Vågor eller partiklar: gammal stridsfråga.

Heisenberg varm vän av Skandinavien

Schrödinger återvänder ej till Tyskland.

Japan har kallat hem ambassadören i Moskva

Det hålles för troligt att han ej återvänder.
Tre divisioner ha beordrats till Korea.

LÄGET YTTERST SPÄNT

Fig. 4.11 The first page of the daily Swedish newspaper *Svenska Dagbladet* the day after the Royal Swedish Academy of Sciences had decided on the physics prize. One can note that the Swedish Academy, also on November 9, 1933, decided to award the literature prize to Ivan Brunin. (Used with permission.)

English version on *nobelprize.org* is the one quoted above. The joint citation for Schrödinger and Dirac with handwritten modifications are difficult to read, so it was simplified to read, in its English version, "for the discovery of new productive forms of atomic theory." Figure 4.11 shows how the prizes were reported on the front page of the daily Swedish newspaper *Svenska Dagbladet.*

The presentation speech in connection with the award ceremony was given by Henning Pleijel, who chaired the Nobel Committee in 1933. An English translation of his presentation speech is available at *nobelprize.org*. He starts with Planck in the year 1900 and works all the way to 1925/26, also mentioning Einstein, Bohr and de Broglie.

Schrödinger is mentioned first: "I will first of all dwell upon Schrödinger's contribution since it is more closely than the others connected to the state of the development which atomic physics had attained at that period of time, particularly as a result of de Broglie's abovementioned theory of matter-waves." Even to this day, a common starting point when teaching quantum mechanics is to start with the Schrödinger equation.

The allotropic forms of molecular hydrogen were mentioned when Pleijel spoke about Heisenberg, but the uncertainty principle that bears Heisenberg's name received greater emphasis — which was, of course, logical.

Dirac, for his part, had formulated the relativistic wave equation for the electron, which yielded electron spin as a natural result — something that had previously been introduced only as a hypothesis to fit experimental data. The greater challenge lay in the equation's prediction of a positively charged electron, which at the time had never been observed. Pleijel summarized the situation memorably:

> This difficulty which at first opposed the theory has now become a brilliant confirmation of its validity. For later on, positive electrons, the positrons, whose existence was stipulated in Dirac's theoretical investigation, have been found by experiment.

Not even Oseen could have asked for a more perfect vindication.

Heisenberg delivered his Nobel Lecture on December 11 in German, while Schrödinger and Dirac gave their lectures the following day; Schrödinger also spoke in his native German (*Les Prix Nobel* 1933). The three lectures differed markedly in character, with Dirac's being the most forward-looking. In it, he predicted the existence of negative protons — and he was, of course, correct.

In the late 1950s, the Nobel Prizes recognized the discovery of the antiproton. In 1958, Pavel Cherenkov, Ilja Frank and Igor Tamm were awarded the prize "for the discovery and the interpretation of the Cherenkov effect," and in 1959, Emilio Segré and Owen Chamberlain received the prize "for their discovery of the antiproton." The Cherenkov effect had played a decisive role in that discovery.

What Dirac could not predict was the wide range of applications his equations have today, as discussed by me and Alexander Balatsky in *Physics Today* (2019).

1932
Heisenberg, Werner
1901 -
Tyskland
"För uppställande av kvantmekaniken, vars användning bl.a. lett till upptäckten av vätets allotropa former".

1933
Schrödinger, Erwin
1887 - 1961
Tyskland
tillsammans med Dirac, Paul Adrien Maurice
"För upptäckandet av nya fruktbärande former av atomteorien".

1933
Dirac, Paul Adrien Maurice
1902 -
England
tillsammans med Schrödinger, Erwin
"För upptäckandet av nya fruktbärande former av atomteorien".

Chapter 5

THE LATER QUANTUM MECHANICS NOBEL PRIZES OF 1945 AND 1954

5.1 The 1945 Nobel Prize to Wolfgang Pauli

Wolfgang Pauli was nominated in 1933 by Oseen. A special report was prepared, also by Oseen, but when the Committee met to make its final decision, Pauli was not among the laureates. Heisenberg and Schrödinger had been discussed many times and were the obvious choices, and with the unexpected discovery of the positron, Dirac had achieved a success that Pauli could not match. Oseen's special report from March 1933 ends with a summary that in translation reads:

> When Professor Thirring, in 1917, was serving as examiner at a school-leaving (maturity) examination in Vienna, he was astonished when one of the candidates stepped forward and told him that he could not entirely accept Thirring's latest work on the theory of relativity. That young man was the then 17-year-old Pauli.
>
> The great ability to penetrate even into difficult and inaccessible fields, of which Pauli gave such a splendid proof at an early age, has remained with him during the sixteen years that have passed since. He is a highly valued author of comprehensive surveys of current research areas (relativity theory, atomic physics). It has sometimes been said that

> Pauli's receptiveness exceeds his originality. Even if this judgment is correct, Pauli has nevertheless, in my view, been one of those who has contributed most effectively to the construction of modern atomic physics.
>
> Uppsala, 1 March 1933
> C. W. Oseen

In his special report, Oseen identified Pauli's most important contribution as the Pauli exclusion principle (Pauli, 1925). He later reiterated a brief statement about Pauli in his special report on Max Born, which will be discussed in Section 5.6 in connection with the 1954 Nobel Prize. Suffice it to say here that Oseen — by then working at his new Nobel Institute in Stockholm — could not endorse yet another Nobel Prize for a discovery that had, in his view, already been recognized through two previous awards.

In 1934, no Nobel Prize in Physics was awarded, nor was it reserved. The prize money was transferred to a special fund, which most likely helped Manne Siegbahn establish and support what came to be known as the Nobel Institute for Physics. The use of such special funds by members of the Nobel Committees of the Royal Swedish Academy of Sciences has been discussed in some detail by Friedman (2001).

5.2 The Nobel Committee, 1934–1945

In Section 4.13, the changes in the Nobel Committee during the 1920s are described. The changes between 1934 and 1945 were much smaller. Vilhelm Carlheim-Gyllensköld passed away in 1934 and was replaced by Axel Lindh, an experimental physicist from Uppsala, in 1935. Oseen kept the grip of theoretical physics on the Committee until he passed away in 1944 and was replaced by Ivar Waller.

5.3 The Pauli Exclusion Principle

In its original form, Pauli realized that within an atomic system, two electrons cannot share the same set of four quantum numbers. This was not an ad hoc assumption limited to electrons, but a direct consequence of the rules governing the construction of two-particle wave functions — rules that apply to all identical fermions. Just as Dirac's relativistic wave equation led to consequences he could scarcely have anticipated, the same was true for Pauli's exclusion principle.

5.4 Nominations and Special Reports

Whereas Oseen was the first to nominate Pauli, he was certainly not the last. The following nominations came in the following years.

1934: One nomination, by Dutch physicist Dirk Coster.
1935: Four nominations, including one from Nobel laureate Max von Laue and one from Nobel laureate Max Planck. The nomination by Leon Brillouin of both Pauli and Samuel Goudsmit makes it clear that the nomination concerned electron spin, first proposed by Goudsmit and George Uhlenbeck.[a]
1936: No nominations.
1937: Five nominations, including one by Hans Thirring, who met Pauli when Pauli was only 17 years old, as follows from the extract of Oseen's special report in 1933, given above.
1938: One nomination, by Schrödinger.
1939: No nominations.
1940: Three nominations, including one by Uhlenbeck (see footnote), the first of a total of four by Greger Wentzel (one of the early pioneers of quantum mechanics and a friend of Pauli), and the second by Dirk Coster.
1941: One nomination (Wentzel).
1942: One nomination, by Ernest Barker from Ann Arbor, Michigan.
1943: Two nominations, one by Wentzel and one by Edgar Meyer, based in Zurich; this was his sole nomination.
1944: One nomination (Wentzel).

This was the situation in 1945, when Einstein submitted his famous nomination by telegram. He wrote it in English, and it reached the Nobel Committee on January 19, 1945. As we have seen for other nominations by Einstein, it was a bit unusual for him to nominate in January.

[a]Over many years Goudsmit and Uhlenbeck received 70 nominations each, of which the one in 1935 was the first for Goudsmit. They never received the Nobel Prize despite their discovery of electron spin. The reason why an award never happened would be worth studying in the archive, but is beyond the scope of this book.

The telegram reads: "nominate wolfgang pauli for the physics-prize stop his contribution to modern quantum theory consisting in so called pauli or exclusion principle prize became fundamental part of modern quantum-physics being independent from the other basic axioms of that theory stop albert Einstein."

In his comprehensive biography of Pauli, Enz (2002) suggests that this was the decisive nomination for the Nobel Committee — and I can only agree. Curiously, six nominations arrived too late, including those from Nobel laureates de Broglie and Franck. In accordance with common practice, these late submissions were classified in the nomination archive as nominations for 1946.

Oseen's special report is discussed above. Oseen made a short statement about Pauli in his special report about Max Born in 1934 (Section 5.6). The next special report was written by Waller in 1945. He was as careful as his mentor, and of course had access to more material to discuss than had Oseen. The proximity to the prizes in 1932/33 was no longer a problem, and the importance of the exclusion principle was now better understood than in 1934. Waller discussed, in addition to the exclusion principle, the connection between spin and Fermi–Dirac statistics (half-integer spin) and Bose–Einstein statistics (integer spin). The important paper in 1940 by Pauli was explicitly pointed out by Waller (Pauli, 1940). Keeping in mind that Pauli, at a young age in 1921, wrote his famous *Encyclpädie Article* on the relativity theory, the final words in his 1940 article are noteworthy:

> In conclusion we wish to state, that according to our opinion the connection between spin and statistics is one of the most important applications of the special relativity theory.

Waller was clear in his recommendation to award Pauli, and the Nobel Committee was unanimous in their decision on September 10, 1945. The citation reads, in Swedish, "för upptäckten av uteslutningsprincipen, även benämnd Pauliprincipen" ("for the discovery of the Exclusion Pricniple, also called the Pauli Principle").

Pauli's Nobel Prize was the first to be awarded after the end of the Second World War. No prize was awarded in 1940, 1941 and 1942. The 1943 prize was reserved for the next year, and in 1944 Otto Stern was awarded the reserved prize and Isidor Isaac Rabi was awarded the

1944 prize. The war made it impossible for either of them to come to Stockholm. At the time of the award announcement on November 15, 1945 (see Fig. 5.1), Pauli was in the United States, at the Institute for Advanced Study in Princeton, and was considering traveling to Stockholm for the Nobel ceremony on December 10, 1945. As cited by Enz (2002, p. 393), Pauli later explained:

> "[I] finally decided to postpone my participation in the ceremony in Stockholm to the next year, after having heard that Stern and Rabi are doing the same."

Instead, the Institute for Advanced Study organized a grand substitute celebration, marking the occasion in style. Pauli's Nobel Prize was the first ever awarded to a resident member of the Institute (Enz, 2002).

On December 10, 1945, Ivar Waller gave the presentation speech in Swedish (*Les Prix Nobel* 1945), which was entirely focused on the exclusion principle. In Pauli's absence, the chargé d'affaires of the Legation of the United States received the Nobel Prize on behalf of Pauli.

December 1946 became a busy period in terms of Nobel lectures. Percy Bridgman was awarded the prize of 1946 for his work on high-pressure physics and gave his Nobel lecture on December 11, 1946. This was followed on December 12 by Stern and finally by Pauli on December 13 (for some reason, Rabi never delivered a Nobel lecture). Pauli gave his lecture in English and supplied it with references, a very unusual procedure for Nobel lectures. It gives his view of theoretical physics as of 1946.

Finally, it is interesting to note that Pauli's award in 1945 was the first in theoretical physics since 1933. Thus, the trio of 1933 became a quartet in 1945: Heisenberg, Schrödinger, Dirac and Pauli. No one could argue that Pauli did not meet the standards of the other three.

5.5 The Nobel Committee, 1947–1954

Only small changes occurred in the Nobel Committee during this period. Henning Pleijel retired from the Committee in 1947 and was replaced by Gustaf Ising from Stockholms Högskola (later Stockholm University), a pioneer in accelerator physics.

Gustaf Ising stayed on the Committee until 1953, and in 1954 Oskar Klein became a member. The monopoly on theoretical physics that Oseen

DAGENS NYHETER.

Nr 312 Upplaga A I dag 36 sidor Fredagen den 16 November 1945 Stockholmsuppl. nr 26976

DRAMATISK UPPTAKT I HELSINGFORS, DEMONSTRANTER SKINGRADE

Mannerheim och militärledningen i skottlinjen redan från början

KLART I USA: ATOMHEMLIGHETEN BEVARAS TILLS GARANTIER VUNNITS

Nobelpris till chilensk diktarinna, atomforskare och foderkonservator

AIV-metodens skapare belönas, två kärnklyvare får var sitt pris

Motylen friges 25 november, däckbristen bromsar trafiken

Syjunta i Örby på 15 damer vann 100.000

Mistral beundrar Selma Lagerlöf

Kosmisk partikel har fångat Pauli

Priset sporrar yngre forskare

SJ:s kolragnar i Polen fredag

Stort ordensregn vid extrakapitel

Lotterivinster i ordningsföljd

Handelskontakt Sverige–Ungern

Flyktingar åter till Österrike

Fig. 5.1 The first page of the daily Swedish newspaper *Dagens Nyheter* on November 16, 1945. The headline for Pauli reads, in translation, "Cosmic particle has captured Pauli." It is also clear from the text below the headline that Pauli has spent his time at the Institute for Advanced Study in Princeton since 1940. The cosmic particle that captured Pauli's interest was the meson (muon), first predicted theoretically by Hideki Yukawa, something for which he was awarded the 1949 Nobel Prize. In the article on the last page of *Dagens Nyheter*, Pauli makes clear that he was never involved in the Manhattan Project (atomic bomb). Used with permission.

had been careful to keep was finally broken with both Waller and Klein selected as members of the Committee.

In 1954, Erik Rudberg from the Metallographic Institute in Stockholm and Gudmund Borelius from the Royal Institute of Technology (KTH) in Stockholm served as adjoint members of the Nobel Committee.

5.6 The 1954 Nobel Prize to Max Born

The award to Wolfgang Pauli in 1945 had the unexpected (I would assume) effect that the physics community realized that maybe quantum mechanics was not exhausted in terms of Nobel Prizes.

By 1946, Max Born had received a total of three nominations, in 1930, 1934 and 1939. The myth that Einstein promoted Born continues to persist even today. In his book *Helgoland* (2022, Swedish translation), Carlo Rovelli comes back twice to how Einstein nominated Born and Pascual Jordan together with Heisenberg for the Nobel Prize. As I showed in Section 4.9, this is incorrect. Einstein never nominated Born or Jordan, although in a handwritten draft nomination, he briefly entertained the idea. By 1946, however, the situation began to change for Born. That year, he was nominated by de Broglie, and the following year by another Nobel laureate, James Franck. This marked the beginning of a surge in nominations.

Figure 5.2 illustrates how Pauli's Nobel Prize in 1945 triggered an explosion of nominations for Born, in sharp contrast to the few scattered nominations he had received during the 1930s. In addition to de Broglie and Franck (who nominated him three times), Born was also nominated by Nobel laureates Fermi (twice) and Bohr (twice).

A clear nomination pressure was building, but the question remained, *For what* should Born receive the Nobel Prize? To answer this, we must return to 1934, when Oseen wrote a long and detailed special report on Born. His conclusion was that Born's theoretical work on crystals did not meet the high standard required for a Nobel Prize. When Tullio Levi-Civita nominated Born in 1939, Oseen responded with a statement that he saw no reason to revise his 1934 assessment. In that earlier report, Oseen had also argued that it would be unfair to Jordan if Born were awarded alone for the work they had done jointly with Heisenberg (Born and Jordan, 1925; Born, Heisenberg and Jordan, 1926). Since Jordan was not nominated in 1934, an award shared between Born and Jordan was not possible. Oseen's principal argument against awarding Born, however, was that quantum mechanics had already been recognized with two Nobel

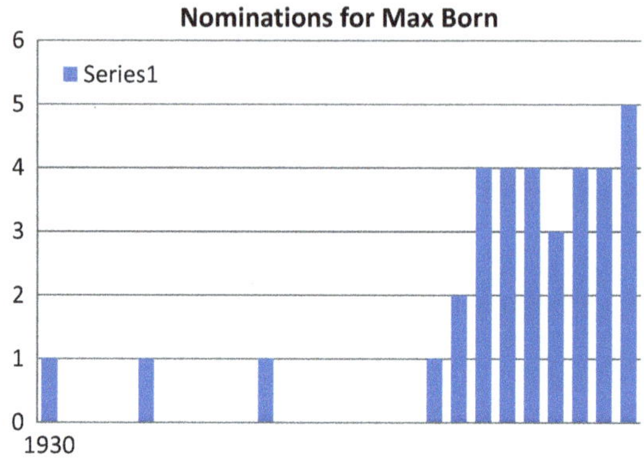

Fig. 5.2 The number of nominations for Max Born as a function of time. The first nomination after the few scattered ones in the 1930s came in 1946, from de Broglie (his nomination had in fact arrived in 1945 after the deadline and was therefore recorded as 1946). The final and highest peak occurred in 1954 — the year Born received the Nobel Prize. This represents a unique nomination pattern.

Prizes, and a third award for a discovery already twice honored was, in his view, out of the question.

The nomination by de Broglie in 1946 (a past-deadline nomination from 1945, recorded as 1946) emphasized Born's contributions to the theory of crystals. The 1947 nomination by James Franck likewise highlighted Born's work on the lattice structure of ionic crystals, as did that of Czesław Białobrzeski. The Nobel Committee took note of these nominations, but saw no reason to commission a special report.

In 1948, the number of nominations increased to four, now including two Nobel laureates among the nominators. In addition to Franck, Enrico Fermi also submitted a nomination. The focus remained on Born's research on crystal theory. This time, the Committee commissioned a special report from Ivar Waller. Born had announced a few years earlier that he would soon publish a book presenting a revised version of his crystal theory. Waller, therefore, recommended waiting for the book's publication before making any decision. He also observed that Born had extended his theory to include electromagnetic and optical properties. In his 1948 report, Waller noted that the book's publication had been delayed, and his suggestion to postpone any award decision was adopted by the Nobel Committee. (The book would not appear until 1954.)

In 1950, Niels Bohr nominated Born for the first time in a nomination he wrote together with his Danish colleague Christian Møller and Torsten Gustafsson from Sweden. Bohr and Møller repeated the nomination in 1951, but it came too late and was included in the nominations for 1952. Bohr and his colleagues took a different angle. In 1950, Bohr wrote (in translation from Danish to English):

> Furthermore, Professor Born has contributed to this formulation (of a rational quantum mechanics) and especially his original investigation of the interpretation of wave mechanical concepts based on clear statistical principles has been of decisive importance for the later development of quantum mechanics and its extensive application to the description of the most diverse atomic processes.

Born was the sole author of the papers in which he formulated the statistical interpretation of the wave function (Born, 1926 (a, b)); thus, complications involving Jordan did not arise. Yet, despite Bohr's nominations of Born in 1950 and 1952, he was still not regarded as a leading candidate for an award. Even in 1953, Born was only one of several physicists mentioned in the Nobel Committee's report to the Academy.

That year's Nobel Prize was awarded to Fritz Zernike "for his demonstration of the phase contrast method, especially for his invention of the phase contrast method" — one of the few Nobel Prizes based entirely on classical physics.

It is difficult to understand from the archive exactly why the Nobel Committee completely changed its mind in 1954, but it is probably not too bold a guess that Oskar Klein's presence in the Committee made a difference. He had had much closer contact with the birth of quantum mechanics in the 1920s than any other Committee member. Waller was commissioned to write a special report, and it is in the report that he quotes Bohr's 1950 nomination given in English translation above. Waller discusses Born's contributions to the theory of crystals, but it is his contributions to quantum mechanics that made him deserve an award. Waller writes (in translation from Swedish to English):

> Before the last world war, only three proposals were submitted for the award for Born's works. After the war, the number of such proposals, as well as the number of different proposers, has increased significantly.

SVENSKA DAGBLADET

TORSDAGEN DEN 4 NOVEMBER 1954

Annonsering i SvD är nyckeln till stora köpstarka marknader

Hård kamp om USA-senaten trots den demokratiska valsegern

Eisenhower tar tillbaka yttrandet om "kallt krig"

Två Nobelpris till tre forskare

Kemipris till USA Fysik till tyskar

Tre tjuvar avslöjades av orädda skolflickor

Finland föreslås till FN

OSTADIGT

MAX BORN:

72-årig Nobelpristagare trodde hans verk glömts

Högertrafik kostar 215 milj. 1959

Höjd bensinskatt i betalningsplan

Ordföranden säger nej till reform

Medicinhistorisk kongress

LINUS PAULING:

Pojkars kemiexperiment blev framtidsavgörande

WALTHER BOTHE:

Älskar att spela piano Målning "semestersynd"

Välkomaffären inför riksdagen

Anonyma breven granskas åter? Ny HD-inlaga

Våra dagars Sydamerika i stort SvD-reportage

Byråingenjören hade dubbelroll i staten och Skänningefirman

FAMILJENYTT

Fig. 5.3 The first page on November 4, 1954, of the Swedish newspaper *Svenska Dagbladet*, with national coverage. The headline for Max Born reads, in translation, "72-year-old Nobel laureate thought his work was forgotten." The prize was divided into two parts with two different citations, the other part going to Walther Bothe for the coincidence method. This is an example of a divided prize with no connection between the two parts, entirely in accordance with Nobel statutes, but it is a bit unusual. This prize was the last for which Einstein nominated, choosing not his old friend Max Born but Walter Bothe. Used with permission.

> This situation undoubtedly reflects a growing appreciation for the significance of Born's contributions, especially his contributions to the fundamental principles and methods of quantum mechanics.

When the Nobel Committee met on September 20, 1954, to finalize and sign their report to the Royal Swedish Academy of Sciences, the decision to award Born was unanimous. The citation in Swedish reads, "för hans grundläggande arbeten inom kvantmekaniken, särskilt hans statistiska tolkning av vågfunktionen" ("for his fundamental research in quantum mechanics, especially for his statistical interpretation of the wavefunction").

Born obtained one half of the prize, and the other half was awarded to Walther Bothe for the coincidence method. Bothe was nominated by Einstein — the very last nomination Einstein made before passing away in 1955. It follows from Fig. 5.3 that Born was surprised when he, at age 72, was awarded the Nobel Prize.

Max Born came to Stockholm for the award ceremony. Ivar Waller gave the presentation speech on December 10, 1954. He mentioned the work Born had done with Jordan and Heisenberg in 1925/26, and proceeded to Born's statistical interpretation of the wavefunction. The next day, Born gave his Nobel lecture with the title "Die statistische Deutung der Quantenmechanic" (*Les Prix Nobel* 1954). Despite having spent more than 20 years in the United Kingdom, he gave the lecture in his native German language. After all, the work for which he was awarded the Nobel Prize was carried out when he was still in Göttingen. For further reading about Born's statistical interpretation in a broader historical context, the article by Pais (1982b) is recommended.

1945
Pauli, Wolfgang
1900 - 1958
Österrike
"För upptäckten av uteslutningsprincipen, även benämnd Pauliprincipen".

1954
Born, Max
1882 -
England
Halvt pris
"För hans grundläggande arbeten inom kvantmekaniken, särskilt hans statistiska
tolkning av vågfunktionen".

Chapter 6

THE 1963 NOBEL PRIZE TO EUGENE WIGNER

I had decided not to include Eugene Wigner in this book — not because he does not fit the theme. On the contrary, his profound work on symmetries in physics most certainly belongs here. The obstacle was that part of the archival material had remained inaccessible until very recently. Wigner's work is closely connected to the theoretical discovery of parity violation by Tsung-Dao Lee and Chen Ning Yang in 1956, for which they were awarded the 1957 Nobel Prize. According to the rules of the Nobel archive, documents remain closed for fifty years, but if a laureate is still alive, the archive stays closed until the laureate has passed away. T. D. Lee died in 2024, and C. N. Yang passed away on October 18, 2025. Therefore, it is now possible to disclose all relevant material concerning Wigner. The prize awarded to Lee and Yang, and the absence of an award for Chien-Shiung Wu, is discussed in a *Physics World* article (Larsson and Wyss, 2026).

In his 1957 special report on Wigner, Lee and Yang, Oskar Klein devoted about half of the four pages to Wigner. According to Klein, Wigner's three most important papers were published in 1927, 1932 and 1952. In the first paper (1927), Wigner examined how a physical system behaves when all spatial coordinates are inverted, from (x, y, z) to $(-x,-y,-z)$. He showed that the stationary states of a particle system have eigenvalues of either +1 or −1 under this transformation. In the second paper (1932), Wigner investigated what happens when time is reversed in a quantum system, replacing t with $-t$. In the third paper (1952), Wigner and two co-workers, building on Wigner's earlier group-theoretical work, carried out a deeper analysis of the parity concept and its application to elementary particles. Klein concluded that Wigner, more than any other physicist, created the theoretical foundation for understanding symmetry principles — particularly inversion operations (spatial and temporal) — which enabled Lee and Yang to identify the possibility of experimentally testing parity violation in weak interaction. Without any doubt, C. S. Wu played the most important role in the experimental test of parity violation (Wu 1957). Wu was not nominated in 1957, and the experiment published that same year, which in any case precluded an award in the same year ("the preceding year"), even if she had been nominated. Had the Nobel Committee waited one year with an awarding parity nomination, Wu could have been included; she received her first nominations in 1958. What is difficult to understand from the 1957 archive is why Wigner was not included together with Lee and Yang. Klein's strong statements about Wigner would definitely have opened up that possibility. Wigner received seven nominations in 1957, whereas Lee and Yang received only one.

Klein's special report was not the first. Wigner was first nominated in 1953 by his former co-author Gregory Breit at Yale University. The nomination resulted in quite a long special report by Ivar Waller.

Waller's final words in English: "In developing the theory for the use of symmetry principles in quantum mechanics, Wigner has undoubtedly taken a leading position and has thereby made a significant contribution to the advancement of physics. This contribution is, however, essentially methodological in nature, and upon reviewing his many perceptive and profound works, I have not found any discovery of such a character that I can at present recommend it for recognition with a Nobel Prize in Physics."

Wigner was nominated almost every year after 1953: 1955 (1 nomination), 1957 (7), 1958 (2), 1959 (1) and 1960 (2). Finally, in the 1960 report from the Nobel Committee to the Royal Swedish Academy of Sciences, the Committee revealed what had happened in 1957. In English translation, it reads:

> In connection with the 1957 nomination, the undersigned, Klein, carried out a supplementary investigation of Wigner's significance for the discovery concerning the parity laws initiated by Lee and Yang, after which the question of dividing the 1957 prize between Wigner and Lee and Yang was thoroughly discussed in the committee. From the discussion, it emerged that Wigner's work on the parity problem could hardly be considered to have had such importance for the contribution of Lee and Yang that a division of the prize would be justified.

It is hardly a bold guess, based on the special reports by Waller in 1953 and Klein in 1957, who argued for including Wigner and who argued against it. The position of the experimental physicists in the Committee is not known.

In 1961 (1 nomination) and 1962 (4), not much happened. In 1963, Wigner received five nominations, of which the one by Victor Weisskopf apparently had an influence on the Nobel Committee:

> In my view, the granting of the Nobel Prize to Lee and Yang was highly justified because of the tremendous impact their work gave to present day experimental physics. However, Lee's and Yang's work would have been impossible and unthinkable without not only the earlier work by Wigner but also his recent work. In my opinion, therefore, if I am permitted to express it in such a blunt form, a Nobel Prize to Lee and Yang without a Nobel Prize to Wigner appears as an inconsistency which I hope the Nobel Committee will correct by granting a prize to Wigner.

The Nobel Committee asked Waller to write a special report. The concluding remarks by Waller read (in English translation): "As should be clear from this and earlier special reports, Wigner has carried out extraordinarily important and innovative work in the theory of atomic nuclei and elementary particles, and his contributions have been essential to the recent development of these fields of research."

With this special report, the Nobel Committee had no problem putting together an award with one half going to Wigner "for his contributions to the theory of the atomic nucleus and the elementary particles, particularly through the discovery and application of fundamental symmetry principles" and the other half to be shared between Maria Goeppert Mayer and J. Hans D. Jensen "for their discoveries concerning nuclear shell structure."

The presentation speech on December 10, 1963, was given by Ivar Waller. It is amusing to note that the first article by Wigner (1927), cited by Klein in his 1957 special report, was published when Lee and Yang were small children. What is less amusing is that the fourth key person concerning parity violation, C. S. Wu, was never awarded, a neglect that cannot be remedied, because she died in 1997.

1963
U.S.A.

Chapter 7

EPILOGUE

In this book I have attempted to describe the Nobel Prizes awarded the pioneers in quantum mechanics: Max Planck, Albert Einstein, Niels Bohr, Louis de Broglie, Werner Heisenberg, Erwin Schrödinger, Paul A.M. Dirac, Wolfgang Pauli, Max Born and Eugene Wigner. It cannot be argued that quantum mechanics has been mistreated in terms of Nobel Prizes.

The two most important theories developed in the 20th century are the theories of relativity and quantum mechanics. The former was never awarded a Nobel Prize, as we have seen in Section 2.2. This does not mean that relativity was not recognized by nominations. For example, the English astronomer Arthur Eddington was nominated in 1932 by his countryman Joseph Petavel for his work on the mathematical development of the theory of relativity. The Nobel Committee noted Eddington's valuable contributions but concluded that if anyone were to be awarded for the theories of relativity, it should be Einstein. However, Einstein was not nominated in 1932, nor at any time after he was awarded the prize. It seems like the nomination community was satisfied that Einstein had been awarded, even if not for his most important contributions to physics.

In this book the main source of information has been the Nobel archive. There are many books on quantum mechanics, but this is the first complete report on all the Nobel Prizes awarded for quantum mechanics more than 50 years ago, with none of the laureates still alive.

There are, of course, Nobel Prizes concerning quantum mechanics from the past 50 years. Those that are most closely related to the theme of this book, the fundamentals of quantum mechanics, are the experimental prizes awarded in 2022 and 2025, recognizing the profound work of John Stewart Bell and Anthony Leggett (Nobel Prize 2003).

ACKNOWLEDGEMENT

I would like to express my sincere thanks to the Director of and Professor at the Center for History of Science Karl Grandin for his invaluable assistance and many engaging discussions. My appreciation also goes to the archivists at the Center, Anne Miche de Malleray and Maria Asp, for their help in making the Nobel archive material available to me. I am also indebted to Diana Kormos Buchwald for her assistance in clarifying the mysterious Einstein nomination. I would like to express my sincere thanks to John Wettlaufer for the foreword to this book. His expertise in the field and thoughtful insights have added immense value to this work.

BIBLIOGRAPHY

Aitchison, I.J.R., MacManus, D.A. and Snyder, T.M. (2004). Understanding Heisenberg's "Magical" Paper of July 1925: A New Look at the Calculation Details, *Am. J. Phys.* **72**, 1370–1379.

Anderson, C.D. (1932). The Apparent Existence of Easily Deflectable Positives, *Science* **76**, 238–239.

Anderson, C.D. (1933). The Positive Electron, *Phys. Rev.* **43**, 491–494.

Bernstein J. (2005). Max Born and the Quantum Theory, *Am. J. Phys.* **73**, 999–1008.

Berry, M. (1998). P.A.M. Dirac: The Purest Soul in Physics, *Physics World* **11**, 36–40.

Bonhoeffer, K.F. and Harteck, P. (1929a). Experimente über Para- und Orthowasserstoff, *Naturwiss.* **17**, 182.

Bonhoeffer, K.F. and Harteck, P. (1929b). Über Para- und Orthowassersstoff, *Z, Phys. Chem.* **B4**, 113–141.

Born, M. (1926a). Zur Quantenmechanic der Stoßvorgänge, *Z. Physik* **37**, 863–867.

Born, M. (1926b). Quantenmechanic der Stoßvorgänge, *Z. Physik* **38**, 803–827.

Born, M. and Jordan, P. (1925). Zur Quantenmechanik, *Z. Physik* **34**, 858–888.

Born, M., Heisenberg, W. and Jordan, P. (1926). Zur Quantenmechanik II, *Z. Physik* **35**, 557–615.

Carlberg, I. (2019). *Nobel. Den gåtfulle Alfred, hans värld och hans pris* (Nordstedts, Sweden).

Carlberg, I. (2023). *Nobel: the enigmatic Alfred and his prizes* (Nordstedts, Sweden).

Chadwick, J. (1932). The Existence of a Neutron, *Proc. Roy. Soc. A* **136**, 692–708.

Dahn, R. (2025). Demythologizing Quantum History, *Physics Today* **78**, 38–43.

Davisson, C. and Germer, L.H. (1927a). The Scattering of Electrons by a Single Crystal of Nickel, *Nature* **119**, 558–560.

Davisson, C. and Germer, L.H. (1927b). Diffraction of Electrons by a Crystal of Nickel, *Phys. Rev.* **30**, 705–740.

Dennison, D.M. (1927). A Note on the Specific Heat of the Hydrogen Molecule, *Proc. Roy. Soc. A* **115**, 483–486.

Dirac, P.A.M. (1925). The Fundamental Equations of Quantum Mechanics, *Proc. Roy. Soc. A* **109**, 642–653.

Dirac, P.A.M (1927). The Quantum Theory of Emission and Absorption of Radiation, *Proc. Roy. Soc. A* **114**, 243–255.

Dirac, P.A.M (1928a). The Quantum Theory of the Electron, *Proc. Roy. Soc. A* **117**, 610–625.

Dirac, P.A.M. (1928b). The Quantum Theory of the Electron, Part II, *Proc. Roy. Soc. A* **118**, 351–361.

Enz, C.P. (2002). *No Time to Be Brief: A Scientific Biography of Wolfgang Pauli* (Oxford University Press).

Euken, A. (1929). Der Nachweis einer Umwandlung der antisymmetrischeen Wasserstoffmolekülart in die symmetrische, *Naturuwiss.* **17**, 182.

Fedak, W.A. and Prentis, J.J. (2009). The 1925 Born and Jordan Paper "On Quantum Mechanics", *Am. J. Phys.* **77**, 128–139.

Friedman, R.M. (2001). *The Politics of Excellence: Behind the Nobel Prize in Science* (A W.H. Freeman Book, Times Books, Henry Holt and Company, New York)

Grandin, K. (2000). *Ett slags modernism i vetenskapen; Teoretisk fysik i Sverige under 1920-talet (A kind of modernism in science: Theoretical physics in Sweden during the 1920s)*, PhD thesis, Universiterstryckeriet, Ekonomikum, Uppsala.

Grandin, K. (2020). Marie Skłodowska Curie — Between Physics and Chemistry and Two Nobel Prizes, *Historia Scientiarum*, 30, pp. 3–18.

Grandin, K. (2021). The Difficult Task to Award Einstein a Nobel Prize, *Il Nuovo Saggiatore* **37** (1–2), 7–16.

Greenspan, N.T. (2005). *The End of the Certain World* (John Wiley & Sons, West Sussex, England).

Griffiths, D.J. (2005). *Introduction to Quantum Mechanics* (Pearson Education, Inc., New Jersey).

Heisenberg, W. (1925). Über quantentheoretische Umdeutung kinematischer und mekanischer Bezihungen, *Z. Physik* **33**, 879–893.

Heisenberg, W. (1927a). Mehrkörperprobleme und Resonanz in der Quantenmechanic. II, *Z. Physik* **41**, 239–267.

Heisenberg, W. (1927b). Über den anschaulichen Inhalt der quantentheoretischen Kinematik und Mekanik, *Z. Physik* **43**, 172–198.

Heisenberg, W. and Pauli, W. (1929). Zur Quantendynamik der Wellenfelder, *Z. Physik* **56**, 1–61.

Heisenberg, W. (1993). *Delen och helheten* (Hallonquists Bokförlag, Sweden) (German original: *Der Teil und das Ganze; Gespräche im Umkreis der Atomphysik*, Harper & Row, 1971).

MacKinnon, E. (1977). Heisenberg, Models, and the Rise of Matrix Mechanics, *Hist. Stud. Phys. Sci.* **8**, 137–188.

Kragh, H.S. (1990). *Dirac; A Scientific Biography* (Cambridge University Press).

Larsson, M. and Balatsky, A. (2019). Paul Dirac and the Nobel Prize in Physics, *Physics Today* **72**, 46–52.

Larsson, M. and Wyss, R. (2026). Chien-Shiung Wu: How she missed the 1957 Nobel Prize, *Physics World* **39**(3), 41.

Nagel, B. (1982). The Discussion Concerning the Nobel Prize for Max Planck, *Science, Technology & Society in the Time of Alfred Nobel*, Pergamon Press, Oxford, pp. 352–376.

Nauenberg, M. (2016). Max Planck and the Birth of the Quantum Hypothesis, *Am. J. Phys.* **84**, 709–720.

Oppenheimer, J.R. and M.S. Plesset (1933). On the Production of the Positive Electron, *Phys. Rev.* **44**, 53–55.

Pais, A. (1982a). *'Subtle is the Lord...': The Science and Life of Albert Einstein* (Oxford University Press, Oxford).

Pais, A. (1982b). Max Born's Statistical Interpretation of Quantum Mechanics, *Science* **218**, 1193–1198.

Pauli, W. (1925). Über den Zusammenhang des Abschlusses der Elektronengruppen im Atom mit der Komplexstruktur der Spektren, *Z. Physik* **31**, 765–783.

Pauli, W. (1940). The Connnection Between Spin and Statistics, *Phys. Rev.* **58**, 716–722.

Les Prix Nobel 1911, P.A Norstedt & Söner (1912).

Les Prix Nobel 1919–1920, P.A Norstedt & Söner (1922).

Les Prix Nobel 1921–1922, P.A Norstedt & Söner (1923).

Les Prix Nobel 1929, P.A Norstedt & Söner (1930).

Les Prix Nobel 1930, P.A Norstedt & Söner (1931).

Les Prix Nobel 1933, P.A Norstedt & Söner (1933).

Les Prix Nobel 1945, P.A Norstedt & Söner (1947).

Les Prix Nobel 1946, P.A Norstedt & Söner (1948).

Les Prix Nobel 1954, P.A Norstedt & Söner (1955).

Rovelli, C. (2022). *Helgoland* (Translation to Swedish, Fri Tanke, Stockholm).

Schrödinger, E. (1926). Quantisierung als Eigenwertproblem, *Ann. Phys.* **79**, 361–376.

Schrödinger, E. (1928). *Collected Papers on Wave Mechanics* (Blackie & Son Limited, London and Glasgow).

Thomson, G.P. and Reid, A. (1927). Diffraction of Cathode Rays by a Thin Film, *Nature* **119**, 890.

Tomonaga, S.-I. (1997). *The Story of Spin*, T. Oka translation from Japanese (U. Chicago Press).

Wigner, E. (1927). Über die Erhaltungssätze in der Quantenmechanik, *Nachrichten von der Gesellschaft der Wissenschaften zu Göttingen, Math. Klasse* **1927**, 375–338.

Wigner, E. (1932). Über die Operation der Zeitumkehr in der Quantenmechanik, *Nachrichten von der Gesellschaft der Wissenschaften zu Göttingen, Math. Klasse* **1932**, 546–559.

Wick, G.C., Wightman, A.S. and Wigner, E. (1952). The Intrinsic Parity of Elementary Particles, *Phys. Rev.* **88**, 101–105.

Wu, C.S., Ambler, E., Hayward, R.W., Hoppes, D.D. and Hudson, R.P. (1957). Experimental Test of Parity Violation in Beta Decay, *Phys. Rev.* **105**, 1413–1415.

NAME INDEX

APPENDIX 1

Taken with permission from Grandin (2000) and translated to English

Carl Wilhelm Oseen, 1879-1944

There is very limited information on Carl Wilhem Oseen available in the English language. Karl Grandin PhD thesis, which he defended at Uppsala University on November 19, 1999 contains comprehensive information on Oseen, however, most of it is written in Swedish and the thesis is not available online, only in paper format (Grandin, 2000). It contains a summary in English, and Karl Grandin has been most kind to allow me to reproduce the part of the summary that discusses Oseen.

Summary from Grandin (2000) on Carl Wilhelm Oseen

In Chapter Two [of the thesis], we are given examples of C. W. Oseen's scientific work and his ambition to create room for modern theoretical physics in Sweden. Theoretical hydrodynamics was his first line of research. In this connection, his aim was to develop a discipline that was purely Swedish by solely publishing his works in Swedish journals, the only result being that it was not until the outbreak of the First World War that foreign scientists became aware of his work. In their respective countries, Oseen and Ludwig Prandtl built up their own schools of hydrodynamics, with their corresponding theoretical technologies, and with different techniques for the formulation of problems and with which to produce results. Oseen's and his pupils' theory was good for hydrodynamical precision measurements, whereas Prandtl's model was applicable to aerodynamics.

In 1921 Oseen tried to establish a laboratory dedicated to experimentation intended to solve problems concerning theoreticians. This has to be regarded as an attempt to change the conditions for pursuing theoretical physics. But it was not a hydrodynamic laboratory that he wished for as he had set about doing research on the theory of liquid crystals, a field in which he hoped to be able to pursue theoretical physics in a new way. However, the head of the department experimental physics in Uppsala perceived this to be a threat to the hegemony of experimental physics. In his position as head of the department, he felt that he and no one else should be responsible for the quality of the experimental research done in Uppsala.

Oseen was to be given an opportunity to change the way physics was done in Uppsala and through his work on the Nobel committee. With the move of Manne Siegbahn and his research school for X-ray spectroscopy to Uppsala, new possibilities were created for theoreticians and experimentalists to collaborate on a common atomic research agenda. It was this opportunity to pursue such joint research work that tempted Oseen. Such a co-operation was, in his wording, "sound"; it was beneficial to both experimentalists and theoreticians to co-operate.

It is often said that experimentalists have dominated Swedish physics. However, the experimental physics culture has not always - if ever - reigned supreme. Rather, in several instances experimental physicists have been anxious to formulate problems to change the direction of research together with theoretical physicists. The joint seminar given by Oseen and Siegbahn in the first half of the 1920s served its primary purpose, that of putting the spotlight on atomic physics.

One might ask why Oseen took an interest in such a development. His own research was mainly concerned with fundamental problems relevant to hydrodynamics, and it could not have been obvious to him that theoretical atomic physics should have "natural" precedence as a more fundamental problem. His experiences of the difficulties in co-operating with the older experimental physicists ought to have played a role, though. The crucial factor in his positive opinion of atomic physics lies in the possibilities of co-operation and the pooling of resources with the experimentalists in the new field of atomic physics. His opinion in this respect was much influenced by his contact with Bohr.

The interest that Oseen took in relativity theory, which did not offer itself to such direct co-operation, has to be understood in two different respects. Partly relativity theory was a symbol of theoretical physics, as such, and partly it offered an important step forward in the development of atomic physics. In the early days of modern quantum theory, Oseen was able to contribute through his profound criticism. His ambitious mathematical critique, that had been so successful in dealing with the problems of hydrodynamics, could demonstrate good points about and shortcomings of the early quantum theories. These mathematical techniques would prove less successful when it came to making a contribution to the new quantum physics; theoretical developments had moved on too far.

In atomic physics Oseen wanted greater clarity, but when it came to the problems in hydrodynamics, he was happy to deal with them as mathematical problems. The mathematics underlying Prandtl's boundary layers made it crucial to criticise his theory. In contrast, as far as quantum physics was concerned, Oseen hoped for lucid concepts that might lay the foundation for the new theoretical technologies and thus make them more intelligible. This hope was not fulfilled in the sense Oseen wished for during the studied period.

In the two next chapters there are examples of two different paths within the modern theoretical physics, namely the careers of Oskar Klein and Ivar Waller. In some instances, they were occupied with similar problems, but studying them offers examples of two different ways of doing theoretical physics. Klein was occupied with more fundamental problems, whereas Waller dealt more with applied theoretical problems. In the chapter on Klein, we are given examples of how Niels Bohr used Danish neutrality to avoid the allied boycott against German physicists, who were amongst the leading figures in atomic physics at this time. He also counted on the support of Ernest Rutherford in Cambridge, with whom he had worked previously. These political experiences would constitute important experiences for Bohr and, later, would form the policy adopted in his institute. These ideas were also being realised through conferences where international co-operation could be pursued. Bohr's first two non-Danish collaborators, H. A. Kramers and Klein, both came from neutral countries. They tried to influence Swedish physicists to take an interest in Bohr's theories, and they also played a role as links to Bohr's old friend in Sweden, Oseen. It was not obvious that Bohr's ideas would receive a favourable reception in Sweden without any effort. His ideas had to be actively exported. Bohr and Oseen both saw atomic physics as fertile common ground for experimentalists and heoreticians to work on. The co-operation with Kramers and Klein was crucial for Bohr in building up his enterprise; in return he helped them by promoting their careers. Despite this backing, this was a difficult professional time for Klein in a way, because the job market for theoretical physicists was very small.

During the first half of the 1920s Klein stayed with Bohr and Kramers in Copenhagen on several occasions before he went to Ann Arbor in the United States to work. During the second part of the decade, he worked in Copenhagen as a close collaborator with Bohr, and he himself made several important contributions to theoretical atomic physics. Niels Bohr's institute was the acknowledged international centre in this area at that time. It served as a theoretical laboratory where theories could be designed, tested and spread.

To a great extent, Kramers and Bohr moulded Klein as a theoretical physicist. By reviewing Bohr's theories in *Kosmos* Klein was able to get a very good overview of the status of theoretical atomic physics at that time. This, his background from Svante Arrhenius' institute for physical chemistry, and his isolation in Ann Arbor, are crucial factors in the shaping of his way of working, his theoretical technologies. The most obvious example of this is his preoccupation with the five-dimensional theory.

In the fourth chapter, we can see how Ivar Waller was occupied primarily with the problem of the interaction of radiation with matter. This area was crucial to the new physics. Waller provides us with an interesting example of how the theoretical physics that developed in connection with the atomic physics became specialised. His preoccupation with one and the same problem is indeed an example of this development. The theoreticians now were so numerous that a division could be made into methods, ways of working, etc. His own occupation with the applied parts of theoretical physics was needed to connect the atomic theories, that became more and more difficult to understand, with the experiments, that in turn became more and more complex. This 'intermediate' theoretical physics, lying in between theory and experiment, was as important as the more fundamental physics done by Klein.

In the fifth chapter, two aspects of one problem are investigated: what line should be taken in regard to the new physics in Sweden. A direct decision on what line of research was to be adopted was taken in making the professorial appointment in Stockholm in 1930, whereas in the Nobel Prize discussions, more indirect decisions were made. Erik Hulthén's contacts with Bohr are revealing in connection with the appointment of the new professor in Stockholm. He was anxious to get Klein to Stockholm, since he, Klein and Bohr were all part of the same group of physicists. This group was not only kept together by social and personal sympathies, but above all by common theoretical technologies. The ambition. was that they would investigate their joint atomic physics problems together, and that their theoretical and experimental technologies should work together.

The question the experts had to address in Stockholm in 1930 was whether one should choose the traditional research agenda, that is, mathematical physics in the form of hydrodynamics, or whether one should choose the new theoretical physics, that is, theoretical atomic physics. They chose the latter. However, the mathematical physics was not declared unusable just because Oskar Klein was given the professorship at Stockholm Högskola, but new theoretical technologies had thus begun to be established at the university departments. According to most histories of hydrodynamics, it poses no problem that theoretical hydrodynamics is now counted amongst the technical applications and that it is mostly pursued at technical colleges. It is never put forward as a problem that it moved from its central position at the university departments, but this was considered to be a problem at the time. The process does not imply that the value of hydrodynamics was put in question, rather it signifies invigoration of the newly founded research at the technical colleges.

Thus, hydrodynamics was replaced by theoretical atomic physics as the dominant theoretical technology at the university departments. In this process, physics and physics research were granted greater scope or space as the old theoretical technology continued at the technical colleges.

When Klein was appointed professor in Stockholm, modern theoretical physics became well established in Sweden. Klein immediately began to develop a modern department. Besides this, he also made an important. contribution through his popular science writings. Oseen too was anxious. to explain the new physics to his older colleagues, who did not always understand the changes that were going on.

By moving to the Nobel Institute Oseen could spend all his time on research, the history of science and Nobel Prize evaluations. In collaboration with Manne Siegbahn, he could work for even more effort to be put into the field of atomic physics in the 1930s. The most important result, however, was that even more space was devoted to theoretical atomic physics in Sweden as Waller took over the department of mechanics and mathematical physics in Uppsala. When the decision came for the professor after Oseen to be nominated, it was apparent that the field of theoretical atomic physics was what mattered.

APPENDIX 2

The original special report about Albert Einstein written by Carl Wilhelm Oseen (1879-1944), dated Uppsala, August 13, 1922. The report is in Swedish, but a translation should be straight forward by means of AI. It is on page 30 Oseen makes the connection between Einstein's law of the photoelectric effect and Bohr's atomic theory.

The special report is from the Nobel Archives, The Royal Swedish Academy of Sciences, Stockholm.

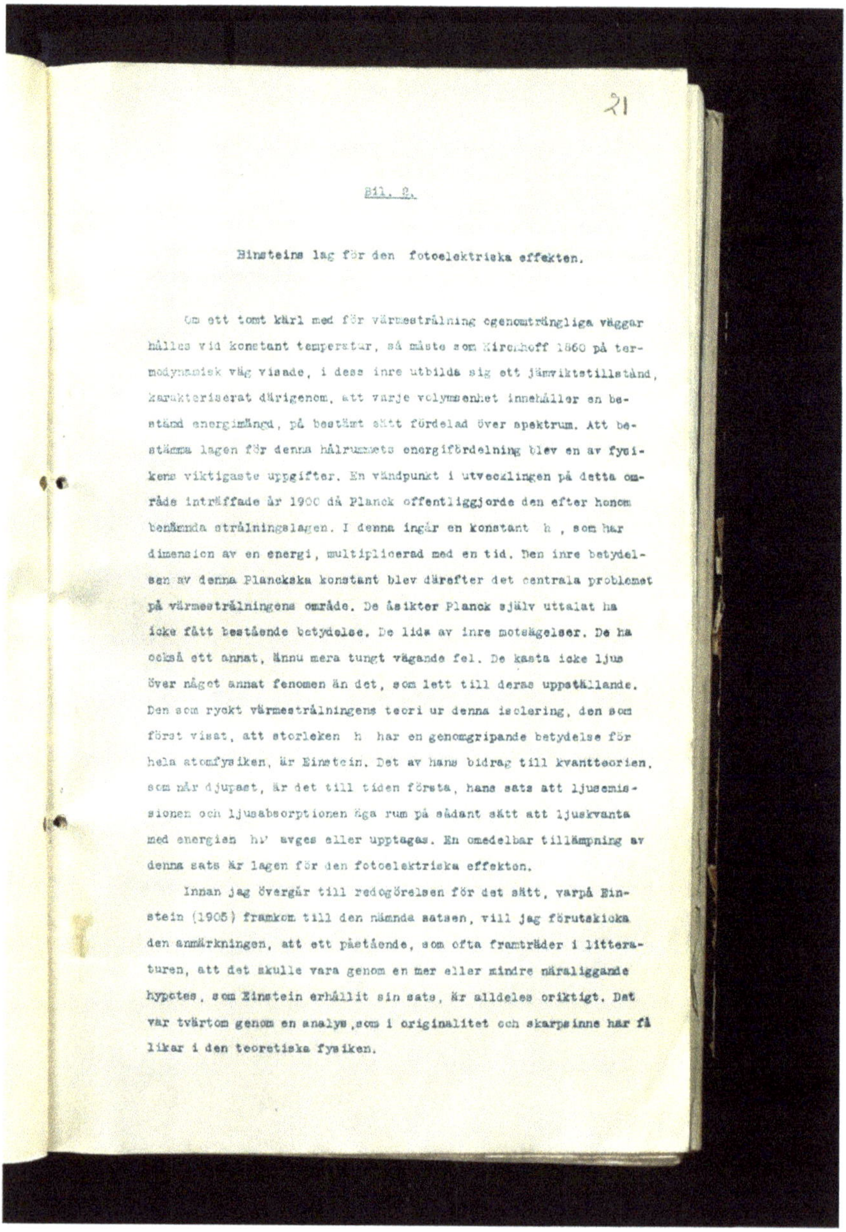

21

Bil. 2.

Einsteins lag för den fotoelektriska effekten.

Om ett tomt kärl med för värmestrålning ogenomträngliga väggar hålles vid konstant temperatur, så måste som Kirchhoff 1860 på termodynamisk väg visade, i dess inre utbilda sig ett jämviktstillstånd, karakteriserat därigenom, att varje volymsenhet innehåller en bestämd energimängd, på bestämt sätt fördelad över spektrum. Att bestämma lagen för denna hålrummets energifördelning blev en av fysikens viktigaste uppgifter. En vändpunkt i utvecklingen på detta område inträffade år 1900 då Planck offentliggjorde den efter honom benämnda strålningslagen. I denna ingår en konstant h , som har dimension av en energi, multiplicerad med en tid. Den inre betydelsen av denna Planckska konstant blev därefter det centrala problemet på värmestrålningens område. De åsikter Planck själv uttalat ha icke fått bestående betydelse. De lida av inre motsägelser. De ha också ett annat, ännu mera tungt vägande fel. De kasta icke ljus över något annat fenomen än det, som lett till deras uppställande. Den som ryckt värmestrålningens teori ur denna isolering, den som först visat, att storleken h har en genomgripande betydelse för hela atomfysiken, är Einstein. Det av hans bidrag till kvantteorien, som når djupast, är det till tiden första, hans sats att ljusemissionen och ljusabsorptionen äga rum på sådant sätt att ljuskvanta med energien hν avges eller upptagas. En omedelbar tillämpning av denna sats är lagen för den fotoelektriska effekten.

Innan jag övergår till redogörelsen för det sätt, varpå Einstein (1905) framkom till den nämnda satsen, vill jag förutskicka den anmärkningen, att ett påstående, som ofta framträder i litteraturen, att det skulle vara genom en mer eller mindre näraliggande hypotes, som Einstein erhållit sin sats, är alldeles oriktigt. Det var tvärtom genom en analys, som i originalitet och skarpsinne har få likar i den teoretiska fysiken.

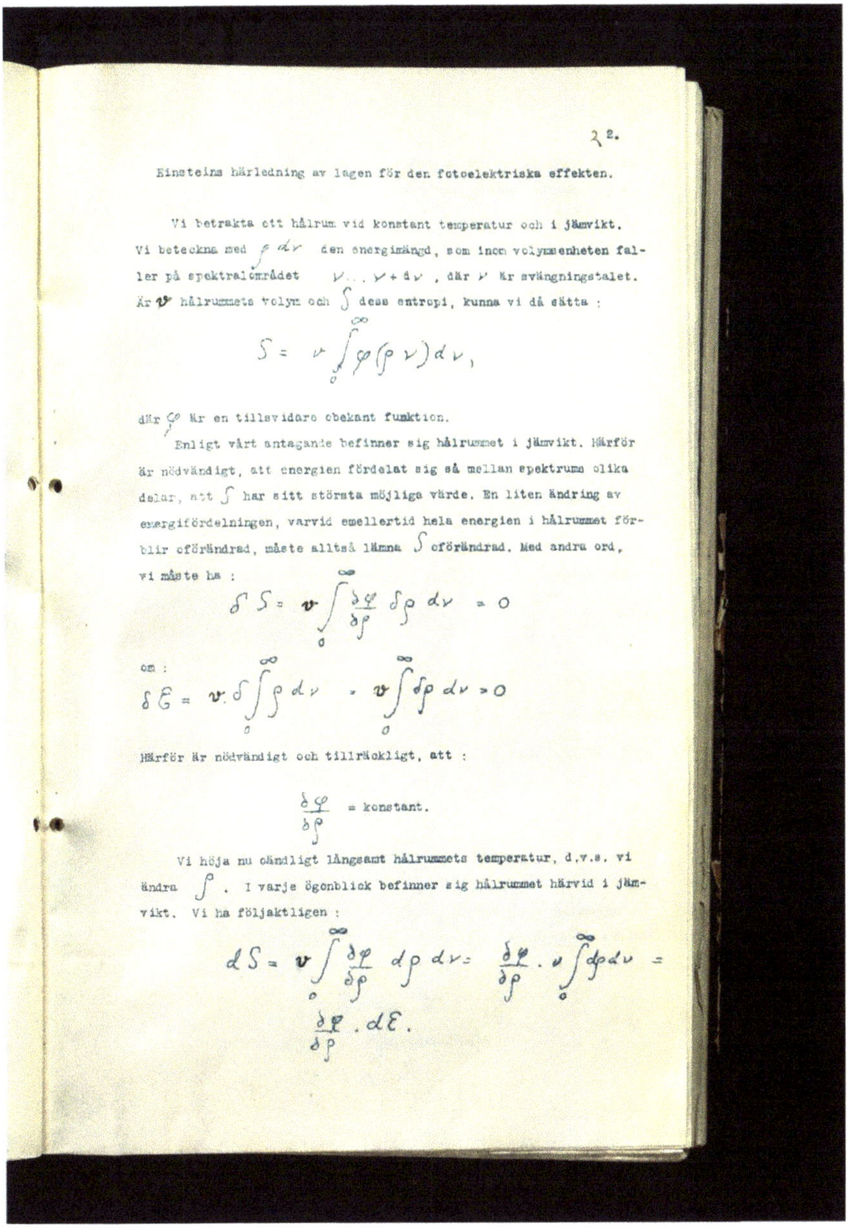

22.

Einsteins härledning av lagen för den fotoelektriska effekten.

Vi betrakta ett hålrum vid konstant temperatur och i jämvikt. Vi beteckna med $\rho\, d\nu$ den energimängd, som inom volymsenheten faller på spektralområdet $\nu \ldots \nu + d\nu$, där ν är svängningstalet. Är v hålrummets volym och S dess entropi, kunna vi då sätta :

$$S = v \int_0^\infty \varphi(\rho\, \nu)\, d\nu,$$

där φ är en tillsvidare obekant funktion.

Enligt vårt antagande befinner sig hålrummet i jämvikt. Härför är nödvändigt, att energien fördelat sig så mellan spektrums olika delar, att S har sitt största möjliga värde. En liten ändring av energifördelningen, varvid emellertid hela energien i hålrummet förblir oförändrad, måste alltså lämna S oförändrad. Med andra ord, vi måste ha :

$$\delta S = v \int_0^\infty \frac{\partial \varphi}{\partial \rho}\, \delta\rho\, d\nu = 0$$

om :

$$\delta E = v\, \delta \int_0^\infty \rho\, d\nu = v \int_0^\infty \delta\rho\, d\nu = 0$$

Härför är nödvändigt och tillräckligt, att :

$$\frac{\partial \varphi}{\partial \rho} = \text{konstant.}$$

Vi höja nu oändligt långsamt hålrummets temperatur, d.v.s. vi ändra ρ. I varje ögonblick befinner sig hålrummet härvid i jämvikt. Vi ha följaktligen :

$$dS = v \int_0^\infty \frac{\partial \varphi}{\partial \rho}\, d\rho\, d\nu = \frac{\partial \varphi}{\partial \rho} \cdot v \int_0^\infty d\rho\, d\nu = \frac{\partial \varphi}{\partial \rho} \cdot dE.$$

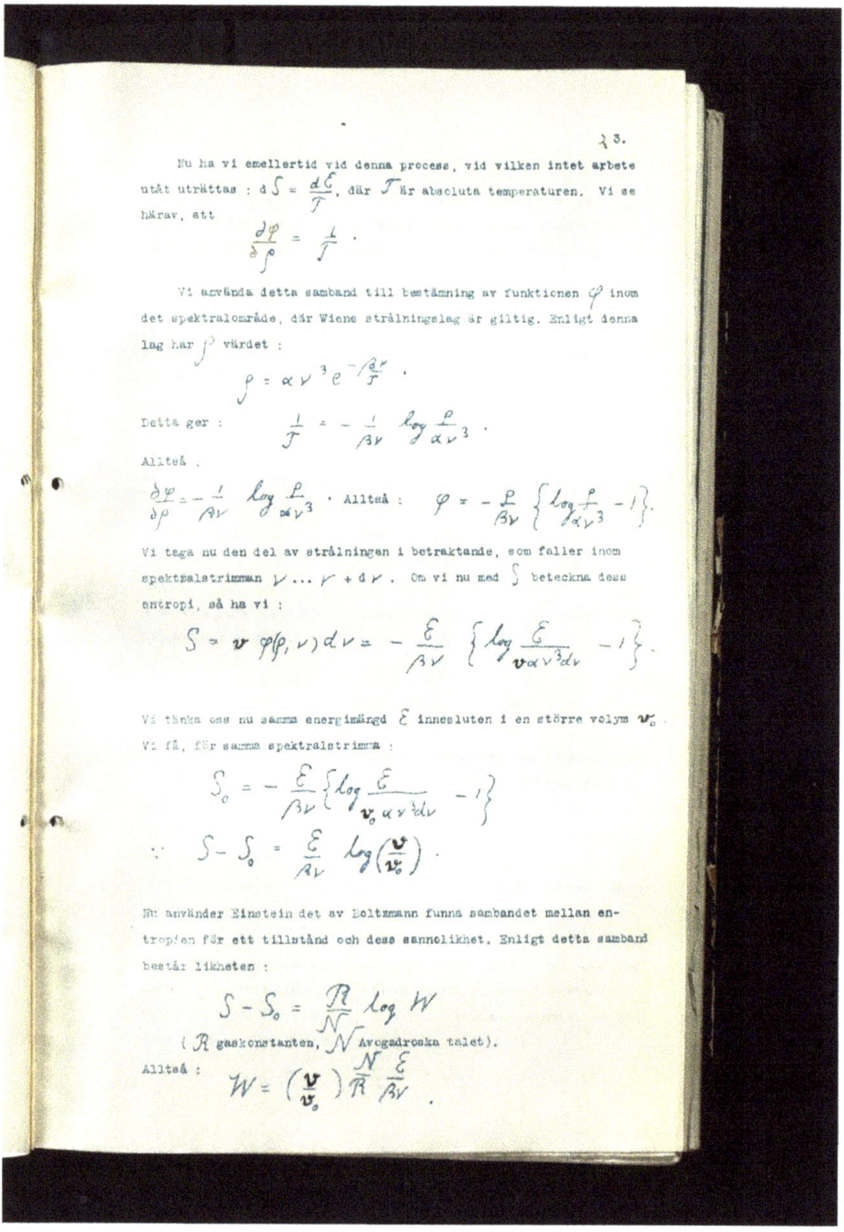

13.

Nu ha vi emellertid vid denna process, vid vilken intet arbete utåt uträttas : $dS = \frac{d\mathcal{E}}{T}$, där T är absoluta temperaturen. Vi se härav, att

$$\frac{\partial \varphi}{\partial \rho} = \frac{1}{T}.$$

Vi använda detta samband till bestämning av funktionen φ inom det spektralområde, där Wiens strålningslag är giltig. Enligt denna lag har ρ värdet :

$$\rho = \alpha \nu^3 e^{-\frac{\beta \nu}{T}}.$$

Detta ger :

$$\frac{1}{T} = -\frac{1}{\beta \nu} \log \frac{\rho}{\alpha \nu^3}.$$

Alltså :

$$\frac{\partial \varphi}{\partial \rho} = -\frac{1}{\beta \nu} \log \frac{\rho}{\alpha \nu^3} \text{. Alltså : } \quad \varphi = -\frac{\rho}{\beta \nu} \left\{ \log \frac{\rho}{\alpha \nu^3} - 1 \right\}.$$

Vi taga nu den del av strålningen i betraktande, som faller inom spektralstrimman $\nu \ldots \nu + d\nu$. Om vi nu med S beteckna dess entropi, så ha vi :

$$S = v\, \varphi(\rho, \nu)\, d\nu = -\frac{\mathcal{E}}{\beta \nu} \left\{ \log \frac{\mathcal{E}}{v \alpha \nu^3 d\nu} - 1 \right\}.$$

Vi tänka oss nu samma energimängd $\mathcal{E}$ innesluten i en större volym v_0. Vi få, för samma spektralstrimma :

$$S_0 = -\frac{\mathcal{E}}{\beta \nu} \left\{ \log \frac{\mathcal{E}}{v_0 \alpha \nu^3 d\nu} - 1 \right\}$$

$$\therefore \quad S - S_0 = \frac{\mathcal{E}}{\beta \nu} \log \left(\frac{v}{v_0} \right).$$

Nu använder Einstein det av Boltzmann funna sambandet mellan entropien för ett tillstånd och dess sannolikhet. Enligt detta samband består likheten :

$$S - S_0 = \frac{R}{N} \log W$$

(R gaskonstanten, N Avogadroska talet).

Alltså :

$$W = \left(\frac{v}{v_0} \right)^{\frac{N}{R} \frac{\mathcal{E}}{\beta \nu}}.$$

24.

Låt oss nu antaga att i en volym v_0 finnas n punktformiga partiklar, fördelade hursomhelst. Vi betrakta en volym v inom v_0 och fråga efter sannolikheten, att de n partiklarna i ett visst ögonblick befinna sig inom denna volym v. Denna sannolikhet är :

$$W = \left(\frac{v}{v_0}\right)^n.$$

Vi se härav : "monokromatisk strålning av ringa täthet (inom giltighetsområdet av Wiens strålningsformel) förhåller sig i värmeteoretiskt avseende så som om den bestode av energikvanta, oberoende av varandra och med storleken $\frac{R\beta\nu}{N}$." Med nu brukligg beteckning är detta $= h\nu$.

Föreställer man sig, att ljusets emission och absorption består i utsändandet eller upptagandet av ett ljuskvantum $h\nu$, får man omedelbart en förklaring av den Stokeska lagen. Tillämpar man samma föreställning på den fotoelektriska effekten, så finner man, att den utsända elektronens energi måste vara lika med det absorberade ljuskvantum $h\nu$, minskat med den energi, som behövs för att frigöra elektronen från atomen och kroppen. Alltså :

$$\frac{1}{2} m v^2 = h\nu - P.$$

Detta är Einsteins lag för den fotoelektriska effekten.

Vid den första Solvay-kongressen i November 1911 avgav Einstein en rapport om det dåvarande läget av problemet om det specifika värmet. I denna rapport meddelar Einstein en ny betraktelse rörande energiutbytet mellan materien och etern, vars resultat sammanfaller med det förut funna. Dess grundtankar äro följande.

Låt oss betrakta en kropp - eller allmännare ett system - med oföränderlig volym, som står i termisk förbindelse med en värmereservoir med oändligt värmeinnehåll och med temperaturen T. Låt $\bar{E}$ vara kroppens medelenergi, $\bar{E} + \varepsilon$ dess energi i ett givet ögonblick och $T + \tau$ dess temperatur i samma ögonblick. Är c systemets specifika värme ha vi då : $\varepsilon = c\,\tau$. Beteckna vi med S entropien för hela det betraktade systemet ha vi vidare :

$$dS = \frac{C\,d\tau}{T+\tau} - \frac{C\,d\tau}{T}$$

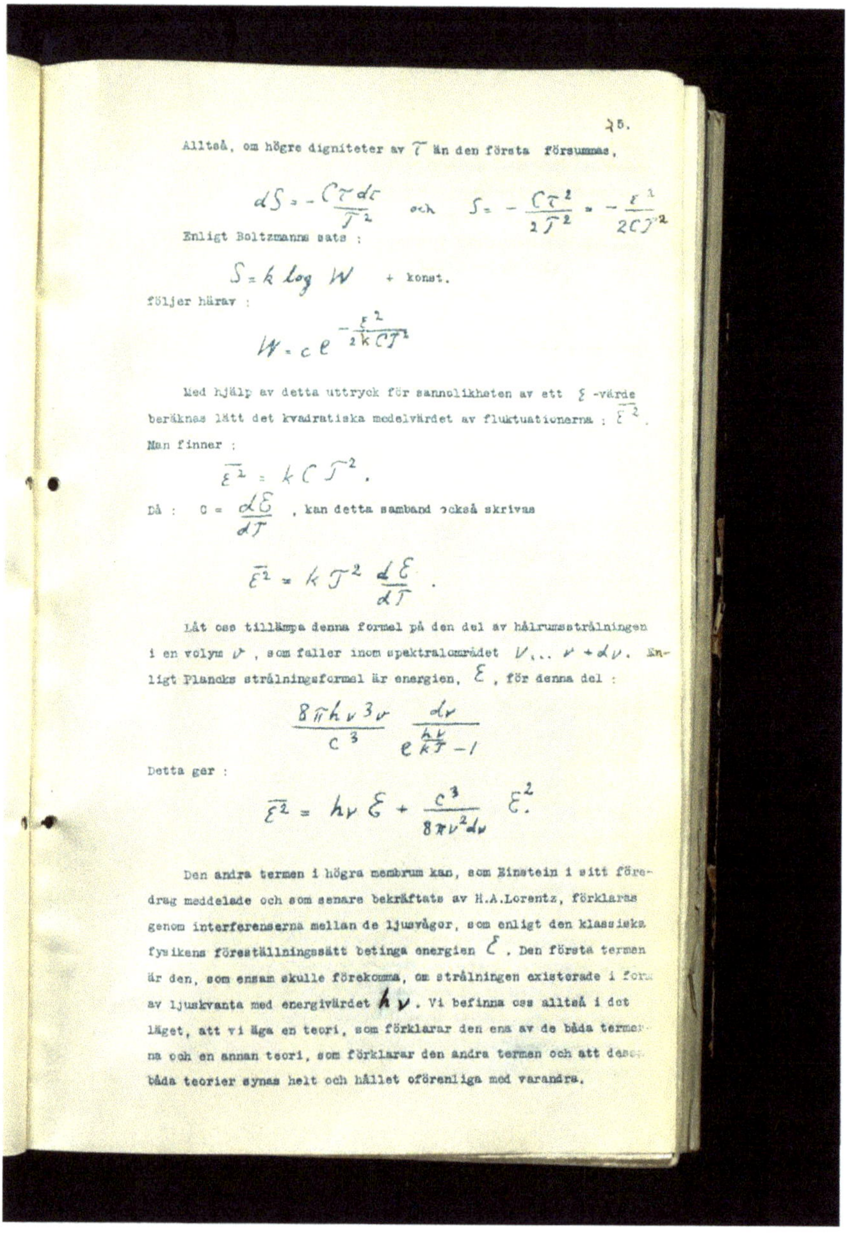

75.

Alltså, om högre digniteter av τ än den första försummas,

$$dS = -\frac{C\tau\, d\tau}{T^2} \quad \text{och} \quad S = -\frac{C\tau^2}{2T^2} = -\frac{\varepsilon^2}{2CT^2}$$

Enligt Boltzmanns sats :

$$S = k \log W + \text{konst.}$$

följer härav :

$$W = c\, e^{-\frac{\varepsilon^2}{2kCT^2}}$$

Med hjälp av detta uttryck för sannolikheten av ett ε-värde beräknas lätt det kvadratiska medelvärdet av fluktuationerna : $\overline{\varepsilon^2}$. Man finner :

$$\overline{\varepsilon^2} = kCT^2.$$

Då : $C = \frac{dE}{dT}$, kan detta samband också skrivas

$$\overline{\varepsilon^2} = kT^2\frac{dE}{dT}.$$

Låt oss tillämpa denna formel på den del av hålrumsstrålningen i en volym v, som faller inom spektralområdet $\nu \ldots \nu + d\nu$. Enligt Plancks strålningsformel är energien, E, för denna del :

$$\frac{8\pi h\nu^3 v}{c^3}\,\frac{d\nu}{e^{\frac{h\nu}{kT}}-1}$$

Detta ger :

$$\overline{\varepsilon^2} = h\nu E + \frac{c^3}{8\pi\nu^2 d\nu}E^2.$$

Den andra termen i högra membrum kan, som Einstein i sitt föredrag meddelade och som senare bekräftats av H.A.Lorentz, förklaras genom interferenserna mellan de ljusvågor, som enligt den klassiska fysikens föreställningssätt betinga energien E. Den första termen är den, som ensam skulle förekomma, om strålningen existerade i form av ljuskvanta med energivärdet $h\nu$. Vi befinna oss alltså i det läget, att vi äga en teori, som förklarar den ena av de båda termerna och en annan teori, som förklarar den andra termen och att dessa båda teorier synas helt och hållet oförenliga med varandra.

26.

Einstein har i sitt förut nämnda föredrag framställt sin tanke i en något annan form än den, som jag för enkelhetens skull valt. I stället för den i kroppen inneslutna energien $\mathcal{E}$ betraktar han den energi, som emitteras eller absorberas av ett ytelement av kroppen. För fluktuationerna av denna energi gälla lagar av samma form som de ovan betraktade. Slutsatserna bli de samma.

Utöver den punkt, som Einstein nådde genom denna sin 1911 offentliggjorda analys, har vetenskapen sedan, beträffande dessa fundamentala frågor, icke nått.

Den experimentella bekräftelsen av Einsteins lag.

Om Einsteins lag för den fotoelektriska effekten är riktig, så är att vänta, att de fotoelektriskt utlösta elektronernas energi skall ha en bestämd övre gräns, nämligen den energi, som tillkommer de elektroner, som utträda ur kroppen utan att genom stötar mot kroppens atomer eller andra elektroner ha förlorat någon energi. Vid en prövning av Einsteins lag är det vid denna högsta energi, som intresset knyter sig. Den benämnes ofta förstenergien.

Det systematiska studiet av förstenergiens beroende av det infallande ljusets svängningstal började år 1907, då E. Ladenburg publicerade en undersökning härav beträffande platina, koppar och zink. Han slöt av sina mätningar, att förstenergien skulle växa proportionellt mot svängningstalets kvadrat. Som Joffé samma år påpekat, var emellertid denna slutsats förhastad, emedan Ladenburg icke tagit hänsyn till alla felkällor. Joffé framhöll, att Ladenburgs mätningar voro förenliga med Einsteins lag. Det var härigenom denna kom in i diskussionen.

Det skulle föra mig för långt att här ge en uttömmande skildring av det experimentella arbete, som nedlagts på prövningen av den Einsteinska lagen. Ett stort antal forskare ha deltagit i detta arbete. Jag nöjer mig med att nämna några av de viktigaste arbetena. - Att förstenergien växer med svängningstalet bekräftades under åren 1908 och 1909 av Ladenburg och Markau, Klages och Hull. Beträffande

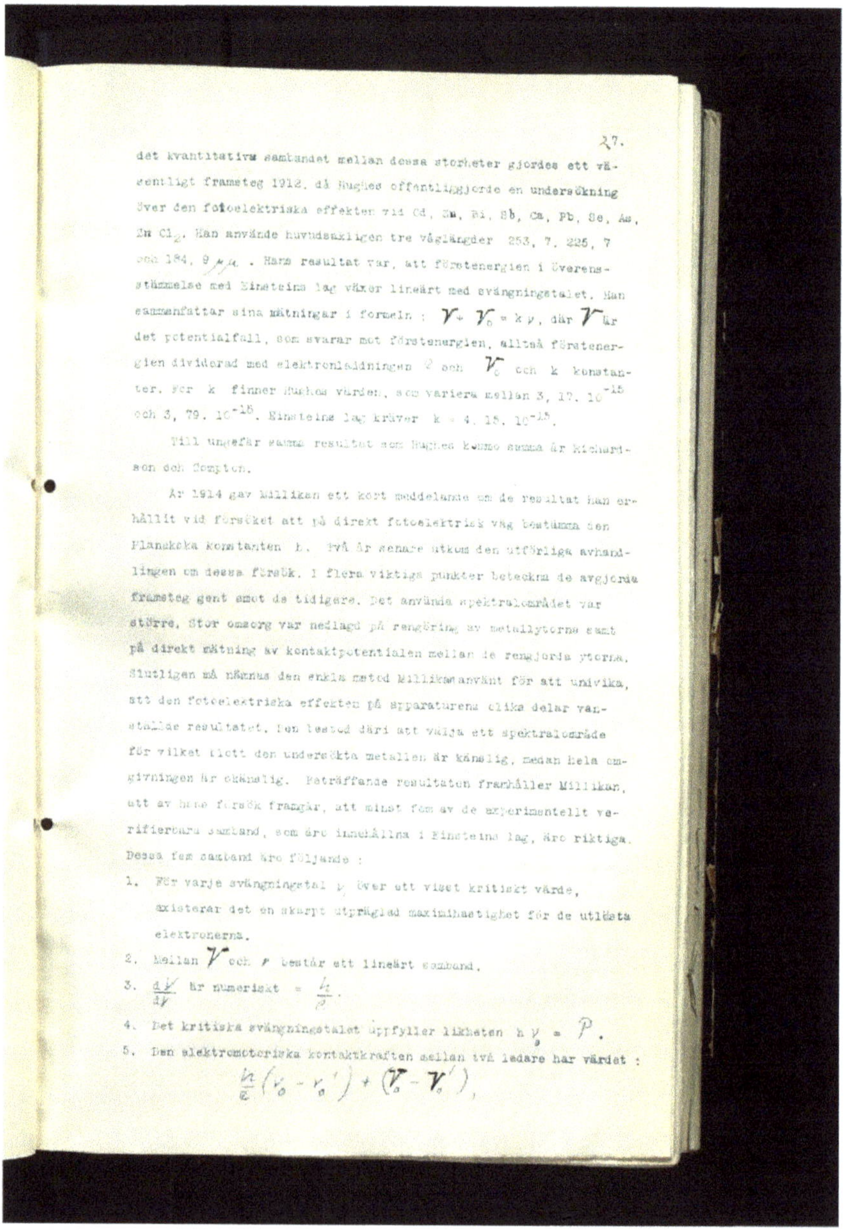

27.

det kvantitativa sambandet mellan dessa storheter gjordes ett väsentligt framsteg 1912, då Hughes offentliggjorde en undersökning över den fotoelektriska effekten vid Cd, Zn, Bi, Sb, Ca, Pb, Se, As, $Zn\ Cl_2$. Han använde huvudsakligen tre våglängder 253, 7, 225, 7 och 184, 9 $\mu\mu$. Hans resultat var, att förstenergien i överensstämmelse med Einsteins lag växer lineärt med svängningstalet. Han sammanfattar sina mätningar i formeln : $V + V_0 = k\,\nu$, där V är det potentialfall, som svarar mot förstenergien, alltså förstenergien dividerad med elektronladdningen e och V_0 och k konstanter. För k finner Hughes värden, som variera mellan $3,17 \cdot 10^{-15}$ och $3,79 \cdot 10^{-15}$. Einsteins lag kräver $k = 4,13 \cdot 10^{-15}$.

Till ungefär samma resultat som Hughes kommo samma år Richardson och Compton.

År 1914 gav Millikan ett kort meddelande om de resultat han erhållit vid försöket att på direkt fotoelektrisk väg bestämma den Planckska konstanten h. Två år senare utkom den utförliga avhandlingen om dessa försök. I flera viktiga punkter betecknade de avgjorda framsteg gent emot de tidigare. Det använda spektralområdet var större. Stor omsorg var nedlagd på rengöring av metallytorna samt på direkt mätning av kontaktpotentialen mellan de rengjorda ytorna. Slutligen må nämnas den enkla metod Millikan använt för att undvika, att den fotoelektriska effekten på apparaturens olika delar vanställde resultatet. Den bestod däri att välja ett spektralområde för vilket blott den undersökta metallen är känslig, medan hela omgivningen är okänslig. Beträffande resultaten framhåller Millikan, att av hans försök framgår, att minst fem av de experimentellt verifierbara samband, som äro innehållna i Einsteins lag, äro riktiga. Dessa fem samband äro följande :

1. För varje svängningstal ν över ett visst kritiskt värde, existerar det en skarpt utpräglad maximihastighet för de utlösta elektronerna.
2. Mellan V och ν består ett lineärt samband.
3. $\frac{dV}{d\nu}$ är numeriskt $= \frac{h}{e}$.
4. Det kritiska svängningstalet uppfyller likheten $h\,\nu_0 = P$.
5. Den elektromotoriska kontaktkraften mellan två ledare har värdet :

$$\frac{h}{e}\left(\nu_0 - \nu_0'\right) + \left(V_0 - V_0'\right),$$

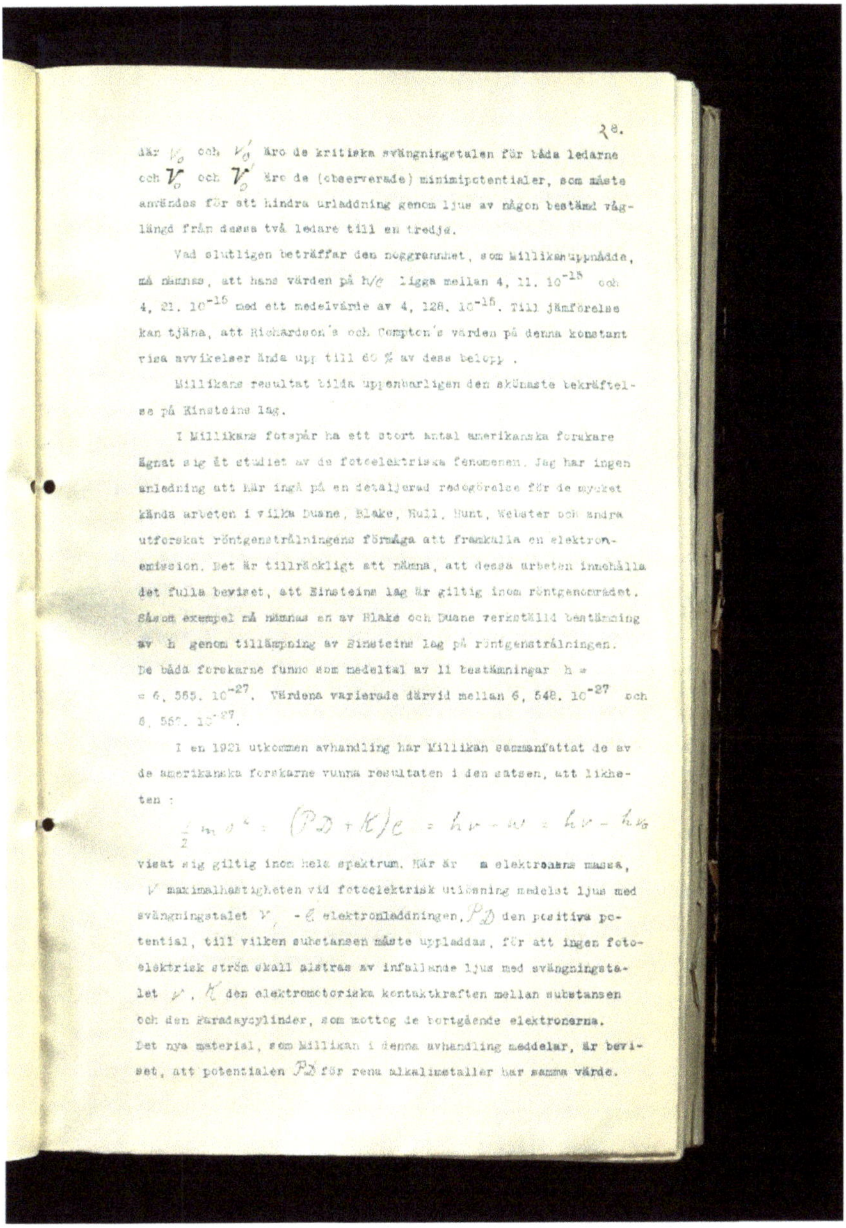

28.

där ν_0 och ν_0' äro de kritiska svängningstalen för båda ledarne och V_0 och V_0' äro de (observerade) minimipotentialer, som måste användas för att hindra urladdning genom ljus av någon bestämd våglängd från dessa två ledare till en tredje.

Vad slutligen beträffar den noggrannhet, som Millikan uppnådde, må nämnas, att hans värden på h/e ligga mellan 4, 11. 10^{-15} och 4, 21. 10^{-15} med ett medelvärde av 4, 128. 10^{-15}. Till jämförelse kan tjäna, att Richardson's och Compton's värden på denna konstant visa avvikelser ända upp till 60 % av dess belopp.

Millikans resultat bilda uppenbarligen den skönaste bekräftelse på Einsteins lag.

I Millikans fotspår ha ett stort antal amerikanska forskare ägnat sig åt studiet av de fotoelektriska fenomenen. Jag har ingen anledning att här ingå på en detaljerad redogörelse för de mycket kända arbeten i vilka Duane, Blake, Hull, Hunt, Webster och andra utforskat röntgenstrålningens förmåga att framkalla en elektronemission. Det är tillräckligt att nämna, att dessa arbeten innehålla det fulla beviset, att Einsteins lag är giltig inom röntgenområdet. Såsom exempel må nämnas en av Blake och Duane verkställd bestämning av h genom tillämpning av Einsteins lag på röntgenstrålningen. De båda forskarne funno som medeltal av 11 bestämningar h = = 6, 555. 10^{-27}. Värdena varierade därvid mellan 6, 548. 10^{-27} och 6, 557. 10^{-27}.

I en 1921 utkommen avhandling har Millikan sammanfattat de av de amerikanska forskarne vunna resultaten i den satsen, att likheten:

$$\frac{1}{2} m v^2 = (PD + K)e = h\nu - w = h\nu - h\nu_0$$

visat sig giltig inom hela spektrum. Här är m elektronens massa, v maximalhastigheten vid fotoelektrisk utlösning medelst ljus med svängningstalet ν, e elektronladdningen, PD den positiva potential, till vilken substansen måste uppladdas, för att ingen fotoelektrisk ström skall alstras av infallande ljus med svängningstalet ν, K den elektromotoriska kontaktkraften mellan substansen och den Faradaycylinder, som mottog de bortgående elektronerna. Det nya material, som Millikan i denna avhandling meddelar, är beviset, att potentialen PD för rena alkalimetaller har samma värde.

29.

Millikan sluter härav, att sannolikt de elektroner, som av det infallande ljuset lösgöres från substansen, äro fria elektroner. Han fremhåller, att om denna slutsats är riktig, förmågan att gripa tag i en elektron och bibringa den energien $h\nu$ måste tillkomma ljuset självt och alltså någon obekant atommekanism därvid icke kan spela någon roll.

Mot de här ovan omtalade resultaten av Millikan kan möjligen en invändning göras. De av Hallwachs ledda undersökningarne över den fotoelektriska effekten förde till den uppfattningen, att den i en metall ockluderade gasen spelar en väsentlig roll för metallens fotoelektriska effekt. Ett starkt stöd fick denna uppfattning i ett arbete av G.Wiedmann och Hallwachs av 1914, i vilket visades, att den fotoelektriska effekten vid kalium är beroende på den av metallen absorberade vätgasen och kan bringas till försvinnande genom rening av metallen från vätgas. Iakttagelser, som peka åt samma håll, finnas i mängd i litteraturen, hos Werner, Welo och andra.

Den förklaring av den ockluderade gasens inverkan, som ligger närmast, är den, att genom växelverkan mellan metallatomerna och gasatomerna vissa elektroner antingen från metallatomerna eller från gasatomerna helt eller delvis lösgöras, och att det är dessa elektroner, som av ljuset föras ut från substansen. Denna uppfattning för till den slutsatsen, att det är storheten P i Einsteins lag :

$$\frac{1}{2}mv^2 = h\nu - P$$

som bör vara beroende av metallens gashalt. Denna slutsats står helt och hållet i överensstämmelse med ett arbete av Millikan och Sonder av 1916. Någon vederläggning av Millikans resultat bilda de förutnämnda arbetena därför icke, om det än måste medgivas, att hela det fotoelektriska fenomenet har en mera komplex karaktär än vad Millikan tidigare förutsatt.

.10.

Den Einsteinska lagen och Bohrs atomteori.

Övertygelsen om den Einsteinska lagens giltighet vilar icke uteslutande och icke ens huvudsakligen på de experimentella bekräftelser, varom ovan varit tal. Redan före offentliggörandet av Millikans bestämning av h var den övertygelsen allmänt spridd, att Einstein med sin djärva lag träffat det rätta. En av de förutsättningar, varpå Bohr uppbyggt sin atomteori, är giltigheten av Einsteins sats, att ljusemissionen och ljusabsorptionen består däri, att en atom avger eller upptar ett kvantum $h\nu$. Nästan alla bekräftelser på Bohrs teori och däribland alla spektroskopiska bekräftelser äro tillika bekräftelser på Einsteins lag. Jag har icke anledning att här ingå på den Bohrska teorien. Jag kan beträffande denna hänvisa till min därom avgivna redogörelse. Jag vill här blott framhålla följande. Den av Bohrs förutsättningar, som till sitt innehåll sammanfaller med Einsteins sats, är det s.k. frekvensvillkoret: $\mathcal{E}_1 - \mathcal{E}_2 = h\nu$. Av alla Bohrs hypoteser är det ingen som så segerrikt bestått anloppen under de nio år, som gått sedan Bohr först uppställde sin teori, som detta frekvensvillkor. Huruvida de vanliga mekaniska lagarna äro giltiga i de stationära tillstånden, veta vi icke. Bohrs regel för bestämning av de stationära banorna har efter Sommerfelds, Epsteins och Schwarzschilds ingripande genomgått en stark utvidgning och omdaning. Beträffande frekvensvillkoret har Bohr själv en gång föreslagit en modifikation i syfte att vinna en teori för Zeemanneffekten. Detta förslag har han emellertid måste återtaga. Samma öde har Sommerfelds teori för "kombinationsdefekter" inom röntgenområdet haft. Den Einsteinska satsen och Bohrs sakligt därmed identiska frekvensvillkor är för närvarande en av de säkraste satser fysiken äger.

Blick på Einsteins verksamhet.

Einsteins talrika arbeten kunna utan tvång sammanföras till fyra olika grupper. En omfattar de arbeten, vilka vila på den förutsättningen, att den klassiska mekaniken är giltig även för molekylerna och atomerna. Dessa arbeten tillhöra det första skedet av Einsteins verksamhet. Deras syfte är antingen att vidga användningsområdet för den kinetiska teorien för materien eller att fördjupa diskussionen av dess allmänna satser. Sommerfeld har föreslagit Einstein till erhållande av Nobelpriset bl.a. för hans fördjupande av Boltzmanns statistiska teori för materien. Det kan möjligen på grund härav vara på sin plats att nämna, att Einstein senare förklarat, att han under denna första period av sin bana icke känt till Gibbs arbeten och att han i annat fall skulle anslutit sig till Gibbs metoder. - Till denna första grupp höra uppenbarligen Einsteins arbeten om den Brownska molekylarrörelsen.

En andra grupp omfattar Einsteins bidrag till kvantteorien. Utom de förut behandlade uppsatserna höra hit hans undersökningar över den fasta kroppens specifika värme samt hans teori av 1916 för värmestrålningen. Beträffande teorien för den fasta kroppens specifika värme må nämnas, att Einstein otvivelaktigt har förtjänsten att ha uttalat den hypotes, som ligger till grund för denna teori, men att vi för utarbetandet av denna teori ha att tacka Debye å ena sidan, Born och v.Kármán å den andra.

En tredje grupp omfattar Einsteins bidrag till den Maxwellska elektromagnetiken. Hit höra de arbeten, som lade grunden för den speciella relativitetsteorien. Särskilt må nämnas upptäckten av hastigheternas additionsteorem samt av sambandet mellan massa och energi. Till denna grupp hör även det tillsammans med Haas utförda arbetet rörande det experimentella påvisandet av de Ampèreska molekylarströmmarne.

Den fjärde gruppen omfattar Einsteins arbeten rörande den allmänna relativitetsteorien.

Inom vart och ett av de områden, som här särskilts, har Einsteins verksamhet satt djupa spår. Att avgiva ett för alla människor och alla tider giltigt omdöme om höjdpunkten av denna intensiva verksamhet är knappast möjligt. För den, som, trots allt vad som under de sista årtiondena framkommit, dock hyser den övertygelsen, att den Newtonska mekaniken måste förbli grundvalen för den fysiska vetenskapen, skall förmodligen den första gruppens arbeten te sig som de betydelsefullaste. För den, som i ett teoretiskt-fysiskt arbete

32.

mest intresserar sig för den matematiska eller kunskapsteoretiska sidan, skall utan tvivel Einsteins allmänna relativitetsteori beteckna höjdpunkten av hans verksamhet. För den åter, som i kvantteorien ser vår tids största bidrag till den teoretiska fysiken, blir det den andra gruppens arbeten, som tilldraga sig det största intresset. För den mätande fysiken har i varje fall ingen annan upptäckt av Einstein ett ens tillnärmelsevis så stort intresse som hans sats om ljusets kvantmässiga emission och absorption.

Sammanfattning.

Vid en tidpunkt, då fysikerna med få undantag ställde sig avvisande mot Plancks kvantteori, har Einstein genom en sällsynt originell och skarpsinnig analys visat, att energiutbytet mellan materien och etern måste ske på det sätt, att en atom avger eller upptar ett energikvantum $h\nu$, där ν är svängningstalet. Såsom tillämpning av denna sats har Einstein uppställt den lagen, att om en elektron fotoelektriskt utlöses från en substans, dess energi efter lösgörandet måste ha värdet $h\nu - P$, där P är det arbete, som behövs för att lösgöra elektronen ur substansen. Denna lag har på det skönaste sätt bekräftats genom mätningar av Millikan och andra. Sin största betydelse och tillika den mest övertygande bekräftelsen har Einsteins sats erhållit därigenom att den är en av de förutsättningar, på vilka Bohr uppbyggt sin atomteori. Nästan alla bekräftelser av Bohrs atomteori äro tillika bekräftelser på Einsteins sats.

Upptäckten av den Einsteinska lagen är utan allt tvivel en av de betydelsefullaste händelserna i fysikens historia. Dess upptäckare synes mig tillfullo förtjäna ett Nobelpris i fysik.

Uppsala den 13 Augusti 1922

C.W. Oseen.

www.ingramcontent.com/pod-product-compliance
Lightning Source LLC
LaVergne TN
LVHW051054210926
848486LV00002B/109
9789819828982